Synthesis Lectures on Engineering, Science, and Technology

The focus of this series is general topics, and applications about, and for, engineers and scientists on a wide array of applications, methods and advances. Most titles cover subjects such as professional development, education, and study skills, as well as basic introductory undergraduate material and other topics appropriate for a broader and less technical audience.

Horst R. Beyer

Quantum Spin and Representations of the Poincaré Group, Part II

With a Focus on Physics and Operator Theory

Springer

Horst R. Beyer
Division of Theoretical Astrophysics
University of Tübingen
Tübingen, Germany

ISSN 2690-0300 ISSN 2690-0327 (electronic)
Synthesis Lectures on Engineering, Science, and Technology
ISBN 978-3-031-95822-9 ISBN 978-3-031-95823-6 (eBook)
https://doi.org/10.1007/978-3-031-95823-6

This Springer imprint is published by the registered company Springer Nature Switzerland AG
The registered company address is: Gewerbestrasse 11, 6330 Cham, Switzerland

Acknowledgments I am indebted to the publisher Springer Nature, in particular, to the Executive Editor Synthesis, Susanne Filler, for the great support. In addition, I am indebted to numerous colleagues, both, from the field of mathematics and physics. The graphics in this text have been created with Wolfram Mathematica® software (www.wolfram.com) and PGF/TikZ software. The text was produced, using the document preparation system LaTeX.

Horst R. Beyer

Competing Interests The author has no competing interests to declare that are relevant to the content of this manuscript.

Introduction

Under changes of the inertial frame of reference, the fields of relativistic quantum field theories must transform under strongly continuous unitary representations of the Poincaré group. The focus of the book is the construction of these representations that provide the basis for the formulation of current relativistic quantum field theories of the scalar fields, the Dirac field and the electromagnetic field.

Such construction is tied to the use of the methods of operator theory that also provide the basis for the formulation of quantum mechanics, up to the interpretation of the measurement process. Also in view of the fact that representation spaces of representations of primary interest in quantum theory are infinite dimensional, the use of these methods is essential. Consequently, the text also calculates the generators of relevant strongly continuous one-parameter groups that are associated with the representations and, wherever appropriate, the corresponding spectrum.

Unlike its counterpart for the scalar field, the representations of the restricted Lorentz group, $\mathcal{L}_+^\uparrow$, associated with the Dirac field are only "partially unitary," in the sense that the one-parameter groups that are associated with Lorentz boosts are not unitary. Hence, their generators are not self-adjoint and, as a consequence, cannot function as "observables" in a one-particle theory. Historically, this fact was one of the reasons to abandon the idea of the development of a "relativistic quantum mechanics," following the model of quantum mechanics, in favor of the quantization of fields. Still, in the process candidates for relativistic one-particle theories emerged that through the process of "second quantization," appear in the construction of corresponding quantum field theories. The text provides the connections of these one-particle theories with the constructed representations.

The representations associated with the Dirac field are induced by a double cover of the restricted Lorentz group $\mathcal{L}_+^\uparrow$, leading to a family of strongly continuous representation of $\mathrm{SL}(2, \mathbb{C})$, the special linear group in 2-dimensions on $\mathbb{C}$, and the concepts of Weyl and Dirac spinor fields.

For the description of non-relativistic spin in quantum mechanics, the book also gives the construction of a double cover of the special unitary group $\mathrm{SU}(2)$, the induced strongly continuous unitary representation of SU(2), calculates the generators and associated spectra corresponding to rotations about the coordinate axes and states the Pauli Interaction Hamiltonian.

The book does not assume any knowledge of Lie groups nor differential geometry. The needed algebraic properties of SO(2), SO(3), SU(2), the Lorentz group $\mathcal{L}$, SL$(2, \mathbb{C})$ and the Poincaré group $\mathcal{P}$ are developed inside the book.

In addition, the book contains a brief rigorous introduction to general matrix Lie groups. We choose for this a direct approach that uses the natural embedding of such a group G into the space $\mathrm{M}(n, \mathbb{K})$, of $n \times n$-matrices over $\mathbb{K} \in \{\mathbb{R}, \mathbb{C}\}$, where n is some non-zero natural number. Viewed from a differential geometric perspective, this text does not go beyond the definition of tangent spaces to matrix Lie groups, although the approach here allows the definition of differentiable vector fields, in particular left invariant vector fields, Lie derivatives, Lie algebras, metrics and related geodesics in matrix Lie groups, essentially leading to a geometric definition of the matrix-valued exponential map. From a perspective of representations of Lie groups in infinite-dimensional Banach spaces such geometric considerations are secondary, although occasionally providing interesting insights.[1] In addition to the algebraic requirement that such a representation R is a homomorphism from a matrix Lie group G into the group $GL(X)$ of bijective bounded linear maps on some Banach space X, R needs to be strongly continuous. Only then R maps one-parameter subgroups or one-parameter semigroups of G to one-parameter subgroups or one-parameter semigroups of $GL(X)$ that have a meaningful generator [20, 24, 29, 44, 4]. Since such generators are generically unbounded and commutators of such operators turned out inconclusive, commutators of unbounded operators do not appear in this text. For this reason, Lie algebras play a minor role in this text.

Finally, in this text and for simplicity, we identify linear maps from $\mathbb{K}^n$ to $\mathbb{K}^n$ with their matrix representation, with respect to the canonical basis of $\mathbb{K}^n$, where $\mathbb{K} \in \{\mathbb{R}, \mathbb{C}\}$ and n is a non-zero natural number. We alert readers from the field of mathematics that, contrary to the standard convention in mathematics, we use in this book the standard convention in physics that scalar products are anti-linear in the first argument and linear in the second argument.

Horst R. Beyer

[1] For instance, geometrical considerations appear in the text, such as in the motivation of the transformation properties of spinor fields and also in the construction of a family of unitary representation of the restricted Lorentz group.

Conventions

For every map f, the symbol $\mathrm{Ran} f$ denotes the set consisting of its assumed values. In particular, if f is a linear map between linear spaces, $\ker f$ denotes the subspace of the domain of f containing those elements that are mapped to the zero vector. For every non-empty set S, the symbol id_S denotes the identity map on S. We always assume the composition of maps (which includes addition, multiplication, etc.) to be maximally defined. For instance, the addition of two maps is defined on the (possibly empty) intersection of their domains.

The symbols $\mathbb{N}$, $\mathbb{R}$, $\mathbb{C}$ denote the natural numbers (including zero), all real numbers and all complex numbers, respectively. The symbols $\mathbb{N}^*$, $\mathbb{R}^*$, $\mathbb{C}^*$ denote the corresponding sets without 0. We call $x \in \mathbb{R}$ positive (negative) if $x \geqslant 0$ ($x \leqslant 0$). We call $x \in \mathbb{R}$ strictly positive (strictly negative) if $x > 0$ ($x < 0$). For every $n \in \mathbb{N}^*$, the symbol S^n denotes unit sphere in Euclidean n-space, i.e., $S^n := \{x \in \mathbb{R}^n : |x| = 1\}$.

For $\mathbb{K} \in \{\mathbb{R}, \mathbb{C}\}$, $n \in \mathbb{N}^*$, $e_1, \ldots, e_n$ denotes the canonical basis of $\mathbb{K}^n$. For every $x \in \mathbb{K}^n$, $|x|$ denotes the canonical norm of x. Further, in connection with matrices, the elements of $\mathbb{K}^n$ are represented as column vectors. The same applies to the elements of direct sums of vector spaces in general. Finally, $M(n \times n, \mathbb{K})$ denotes the vector space of $n \times n$ matrices with entries from $\mathbb{K}$. The $n \times n$ unit matrix is denoted by $E_{n \times n}$ or, if there is no confusion possible, by E. For every $A \in M(n \times n, \mathbb{K})$, $\det(A)$ denotes its determinant and $\mathrm{Tr}(A)$ its trace. To indicate components of vectors, matrices and coordinates, this text uses only lower indices. In particular, we do not use the Einstein summation convention.

For each $k \in \mathbb{N}$, $n \in \mathbb{N}^*$, $\mathbb{K} \in \{\mathbb{R}, \mathbb{C}\}$ and each non-empty open subset Ω of $\mathbb{R}^n$, the symbol $C^k(\Omega, \mathbb{C})$ denotes the linear space of continuous and k-times continuously differentiable complex-valued functions on Ω. Further, $C_0^k(\Omega, \mathbb{C})$ denotes the subspace of $C^k(\Omega, \mathbb{C})$ containing those elements that have a compact support in Ω. If Ω is bounded, $C^k(\overline{\Omega}, \mathbb{C})$ is defined as the subspace of $C^k(\Omega, \mathbb{C})$ consisting of those elements for which there is an extension to an element of $C^k(V, \mathbb{C})$ for some open subset V of $\mathbb{R}^n$ containing $\overline{\Omega}$. The superscript k is omitted if $k = 0$. For every map $f: U \to \mathbb{K}^n$ which is defined on some subset $U \subset \mathbb{R}^n$ as well as differentiable in $x \in U$, $f'(x) \in M(n, \mathbb{K})$ denotes the derivative of f in x defined by

$$f'(x)_{ij} := \frac{\partial f_i}{\partial x_j}(x)$$

for all $i, j \in \{1, \dots, n\}$. In addition, in the case $n = 1$, we define the gradient of f in x by

$$(\nabla f)(x) := \sum_{i=1}^{n} \frac{\partial f}{\partial x_i}(x)\, e_i \, .$$

For a differentiable map γ from some non-trivial open interval I of $\mathbb{R}$ into $\mathbb{K}^n$ or $M(n, \mathbb{K})$, the corresponding derivative is defined component-wise. Further, $BC(\mathbb{R}^n, \mathbb{C})$ denotes the subspace of $C(\mathbb{R}^n, \mathbb{C})$ consisting of those functions which are bounded. $C_\infty(\mathbb{R}^n, \mathbb{C})$ denotes subspace of $C(\mathbb{R}^n, \mathbb{C})$ containing those functions f satisfying

$$\lim_{|x| \to \infty} f(x) = 0 \, .$$

Throughout the course, Lebesgue integration theory is used in the formulation of [50]. Compare also Chap. 3 in [30] and Appendix A in [67]. If not indicated otherwise, the terms "*almost everywhere*" (a.e.), "*measurable*", "*integrable*", etc., refer to the Lebesgue measure v^n on $\mathbb{R}^n$, $n \in \mathbb{N}^*$. The appropriate n will be clear from the context. Nevertheless, we often mimic the notation of the Riemann-integral to improve readability. We follow common usage and don't differentiate between a function f which is almost everywhere defined (with respect to the chosen measure) on some set and the associated equivalence class consisting of all functions which are almost everywhere defined on that set and differ from f only on a set of measure zero. In this sense, for $p \geqslant 1$, the symbol $L^p_{\mathbb{C}}(\Omega, \rho)$, where ρ is some a.e. defined and positive real-valued and locally integrable function on Ω, denotes the vector space of all complex-valued measurable functions f which are defined on Ω and such that $|f|^p$ is integrable with respect to the measure ρv^n. For every such f, we define the L^p-norm $\|f\|_p$ corresponding to f by

$$\|f\|_p := \left(\int_{\Omega} \rho |f|^p \, dv^n \right)^{1/p} .$$

Equipped with $\| \; \|_p$, $L^p_{\mathbb{C}}(\Omega, \rho)$ is a Banach space. In addition, we define in the special case $p = 2$ a scalar product $\langle | \rangle_2$ on $L^2_{\mathbb{C}}(\Omega, \rho)$ by

$$\langle f | g \rangle_2 := \int_{\Omega} \rho f^* g \, dv^n \, ,$$

for all $f, g \in L^2_{\mathbb{C}}(\Omega, \rho)$, where $*$ denotes complex conjugation on $\mathbb{C}$. As a consequence, $\langle | \rangle_2$ is antilinear in the first argument and linear in its second. This convention is going to be used for sesquilinear forms in general. If ρ is constant of value 1, we omit any reference to ρ in the previous symbols. $L^{\infty}_{\mathbb{C}}(\Omega)$ denotes the vector space of complex-valued measurable bounded functions on Ω. For every $f \in L^{\infty}_{\mathbb{C}}(\Omega)$ we define

$$\|f\|_{\infty} := \sup_{x\in\Omega}|f(x)| .$$

Equipped with $\| \, \|_{\infty}$, $L^{\infty}_{\mathbb{C}}(\Omega)$ is a complex Banach space.

Finally, standard results and nomenclature of Operator Theory are used. For this, compare textbooks on Functional Analysis, e.g., [49] Volume I, [50, 74]. In particular, for every non-trivial normed vector space $(X, \| \, \|_X)$ and any normed vector space $(Y, \| \, \|_Y)$ over the same field, we denote by $\mathrm{L}(X, Y)$ the vector space of continuous linear maps from X to Y. Equipped with the operator norm $\| \, \|_{\mathrm{op},X,Y}$, defined by

$$\|A\|_{\mathrm{op},X,Y} := \sup_{\xi\in X, \|\xi\|_X=1} \|A\xi\|_Y$$

for all $A \in \mathrm{L}(X, Y)$, $\mathrm{L}(X, Y)$, is a normed vector space which is complete if $(Y, \| \, \|_Y)$ is complete. Frequently, indices in $\| \, \|_{\mathrm{op},X,Y}$ are omitted if there is no confusion possible. Finally, for every non-empty subset U of some normed vector space $(X, \| \, \|_X)$ and any normed vector space $(Y, \| \, \|_Y)$, the symbol $C(U, Y)$ denotes the vector space of continuous functions from U to Y.

Contents

Spinor Representations of the Poincaré Group 1

In the following, we are going to construct a double cover $\Phi_2 : \mathrm{SL}(2, \mathbb{C}) \to \mathcal{L}_+$ and $\Phi_2 : \mathrm{SL}(2, \mathbb{C}) \to \mathcal{L}_+$ of $\mathcal{L}_+$. Later on Φ_2 is used in the definition of the spinor representations of $\mathrm{SL}(2, \mathbb{C})$ corresponding to spins 1/2 and 1.

1.1 Construction of a Double Cover of $\mathcal{L}_+^\uparrow$

In a first step, we define an analogue of $\sigma : \mathbb{R}^3 \to V_3$, defined in Part I, in the construction of a double cover Φ_1 of SO(3), $\sigma : \mathbb{R}^4 \to \mathrm{H}(2, \mathbb{C})$ by

$$\sigma(x) := \begin{pmatrix} x_0 + x_3 & x_1 - i x_2 \\ x_1 + i x_2 & x_0 - x_3 \end{pmatrix} = x_0\sigma_0 + x_1\sigma_1 + x_2\sigma_2 + x_3\sigma_3$$

$$= x_0\sigma_0 - i\,\sigma({}^t(x_1, x_2, x_3)) \;, \tag{1.1}$$

for every $x \in \mathbb{R}^4$, where σ_0 denotes the 2×2 unit matrix,

$$\sigma_1 := \begin{pmatrix} 0 & 1 \\ 1 & 0 \end{pmatrix} \;, \; \sigma_2 := \begin{pmatrix} 0 & -i \\ i & 0 \end{pmatrix} \;, \; \sigma_3 := \begin{pmatrix} 1 & 0 \\ 0 & -1 \end{pmatrix} \;,$$

are the Pauli spin matrices and $\mathrm{H}(2, \mathbb{C})$ denotes the real subspace of $\mathrm{M}(2, \mathbb{C})$ consisting of all Hermitian matrices.

The map σ is well-defined, since

H. R. Beyer, *Quantum Spin and Representations of the Poincaré Group, Part II*,
Synthesis Lectures on Engineering, Science, and Technology,
https://doi.org/10.1007/978-3-031-95823-6_1

$$(\sigma(x))^* = \begin{pmatrix} x_0 + x_3 & x_1 - ix_2 \\ x_1 + ix_2 & x_0 - x_3 \end{pmatrix} = \sigma(x) \ ,$$

for every $x \in \mathbb{R}^4$.

In addition, we define $\langle \ , \ \rangle : \mathrm{H}(2, \mathbb{C}) \times \mathrm{H}(2, \mathbb{C}) \to \mathbb{R}$ by

$$\begin{aligned} \langle A, B \rangle &:= \frac{1}{2} [\det(A + B) - \det(A) - \det(B)] \\ &= \frac{1}{2} \left(a_{11} b_{22} + a_{22} b_{11} - a_{12} b_{12}^* - a_{12}^* b_{12} \right) \\ &= -\frac{1}{2} [\operatorname{Tr}(AB) - \operatorname{Tr}(A)\operatorname{Tr}(B)] \ , \end{aligned}$$

for all

$$A = \begin{pmatrix} a_{11} & a_{12} \\ a_{12}^* & a_{22} \end{pmatrix} \ , \ B = \begin{pmatrix} b_{11} & b_{12} \\ b_{12}^* & b_{22} \end{pmatrix} \in \mathrm{H}(2, \mathbb{C}) \ .$$

We note that

$$\begin{aligned} \langle A, A \rangle &= \frac{1}{2} [\det(2A) - \det(A) - \det(A)] \\ &= \frac{1}{2} [4 \det(A) - \det(A) - \det(A)] = \det(A) \ , \end{aligned}$$

for every $A \in \mathrm{H}(2, \mathbb{C})$ (Fig. 1.1).

Lemma 1.1 *(i)* $\langle \ , \ \rangle$ *defines a non-degenerate symmetric bilinear form on* $\mathrm{H}(2, \mathbb{C})$.
(ii) $\sigma : \mathbb{R}^4 \to \mathrm{H}(2, \mathbb{C})$ *is an isomorphism between the real vector spaces* $\mathbb{R}^4$ *and* $\mathrm{H}(2, \mathbb{C})$ *such that*

$$\langle \sigma(x), \sigma(y) \rangle = x \cdot y \ ,$$

for all $x, y \in \mathbb{R}^4$.

$$\begin{array}{c} (\mathrm{H}(2, \mathbb{C}), \langle \ , \ \rangle) \\ \uparrow \sigma \\ (\mathbb{R}^4, \cdot) \end{array}$$

Fig. 1.1 The map σ is an isometric isomorphism between the real semi-inner product spaces $(\mathbb{R}^4, \cdot)$ and $(\mathrm{H}(2, \mathbb{C}), \langle \ , \ \rangle)$

Proof For $x, y \in \mathbb{R}^4$ and $\lambda \in \mathbb{R}$, it follows that

$$\sigma(x+y) = \begin{pmatrix} (x+y)_0 + (x+y)_3 & (x+y)_1 - i\,(x+y)_2 \\ (x+y)_1 + i\,(x+y)_2 & (x+y)_0 - (x+y)_3 \end{pmatrix}$$

$$= \begin{pmatrix} x_0 + y_0 + x_3 + y_3 & x_1 + y_1 - ix_2 - iy_2 \\ x_1 + y_1 + ix_2 + iy_2 & x_0 + y_0 - x_3 - y_3 \end{pmatrix} = \sigma(x) + \sigma(y) \ ,$$

and

$$\sigma(\lambda x) = \begin{pmatrix} (\lambda x)_0 + (\lambda x)_3 & (\lambda x)_1 - i\,(\lambda x)_2 \\ (\lambda x)_1 + i\,(\lambda x)_2 & (\lambda x)_0 - (\lambda x)_3 \end{pmatrix} = \lambda\, \sigma(x) \ .$$

In particular, $\sigma(x) = 0$ implies that

$$x_0 + x_3 = x_0 - x_3 = x_1 - ix_2 = x_1 + ix_2 = 0$$

and hence that $x = 0$. As a consequence, σ is a monomorphism. On the other hand, if

$$A = \begin{pmatrix} a_{11} & a_{12} \\ a_{21} & a_{22} \end{pmatrix}$$

is a Hermitian element of $\mathrm{M}(2, \mathbb{C})$, then a_{11}, a_{22} are real, $a_{21} = a_{12}^*$, and hence

$$\sigma\left(\frac{1}{2}\,(a_{11} + a_{22}), \mathrm{Re}(a_{12}), -\mathrm{Im}(a_{12}), \frac{1}{2}\,(a_{11} - a_{22})\right)$$

$$= \begin{pmatrix} \frac{1}{2}\,(a_{11} + a_{22}) + \frac{1}{2}\,(a_{11} - a_{22}) & \mathrm{Re}(a_{12}) + i\mathrm{Im}(a_{12}) \\ \mathrm{Re}(a_{12}) - i\mathrm{Im}(a_{12}) & \frac{1}{2}\,(a_{11} + a_{22}) - \frac{1}{2}\,(a_{11} - a_{22}) \end{pmatrix}$$

$$= \begin{pmatrix} a_{11} & a_{12} \\ a_{12}^* & a_{22} \end{pmatrix} = A \ .$$

As a consequence, σ is also surjective and hence an isomorphism. Further, for $x, y \in \mathbb{R}^4$, it follows that

$$\begin{aligned} \langle \sigma(x), \sigma(y) \rangle &= \frac{1}{2}\big[(x_0 + x_3)(y_0 - y_3) + (x_0 - x_3)(y_0 + y_3) \\ &\quad - (x_1 - ix_2)(y_1 - iy_2)^* - (x_1 - ix_2)^*(y_1 - iy_2)\big] \\ &= \frac{1}{2}\big[2\,(x_0 y_0 - x_3 y_3) - (x_1 - ix_2)(y_1 + iy_2) - (x_1 + ix_2)(y_1 - iy_2)\big] \\ &= x_0 y_0 - x_1 y_1 - x_2 y_2 - x_3 y_3 = x \cdot y \ . \end{aligned}$$

□

For later use, we note that

$$\det(\sigma(x)) = \langle \sigma(x), \sigma(x) \rangle = x \cdot x \ ,$$

for every $x \in \mathbb{R}^4$. In the next step, we define for every $G \in \mathrm{SL}(2, \mathbb{C})$ a corresponding real linear transformation $L_G : \mathrm{H}(2, \mathbb{C}) \to \mathrm{H}(2, \mathbb{C})$ by

$$L_G(A) := G A G^* \ ,$$

for every $A \in \mathrm{H}(2, \mathbb{C})$, where we note that L_G is well-defined, since

$$(G A G^*)^* = G A^* G^* = G A G^* \ ,$$

for every $A \in \mathrm{H}(2, \mathbb{C})$. We note that $L_{-G}(A) = L_G(A)$ and that

$$\begin{aligned}
\langle L_G(A), L_G(B) \rangle &= \frac{1}{2} [\det(L_G(A) + L_G(B)) - \det(L_G(A)) - \det(L_G(B))] \\
&= \frac{1}{2} \left[\det(G A G^* + G B G^*) - \det(G A G^*) - \det(G B G^*) \right] \\
&= \frac{1}{2} \left[\det(G(A + B)G^*) - \det(G A G^*) - \det(G B G^*) \right] \\
&= |\det(G)|^2 \, \frac{1}{2} [\det(A + B) - \det(A) - \det(B)] \\
&= |\det(G)|^2 \, \langle A, B \rangle = \langle A, B \rangle \ ,
\end{aligned}$$

for $A, B \in \mathrm{H}(2, \mathbb{C})$, i.e., we have that $L_{-G} = L_G$ and that L_G is isometric. Hence, $\sigma^{-1} \circ L_G \circ \sigma$ is a Lorentz transformation. Furthermore, we arrive at the following result (Fig. 1.2).

Theorem 1.2 *By*

$$\Phi_2(G) := \sigma^{-1} \circ L_G \circ \sigma$$

for every $G \in \mathrm{SL}(2, \mathbb{C})$, *there is given an epimorphism* $\Phi_2 : \mathrm{SL}(2, \mathbb{C}) \to \mathcal{L}_+^\uparrow$. *In particular,*

$$\Phi_2(G_1) = \Phi_2(G_2)$$

for $G_1, G_2 \in \mathrm{SL}(2, \mathbb{C})$ *if and only if* $G_2 = \pm G_1$.

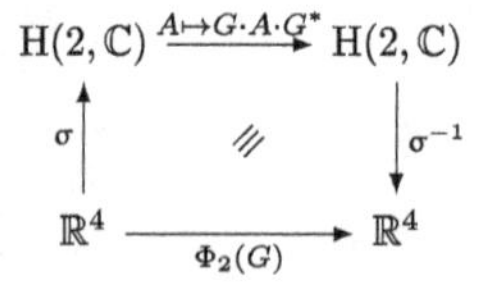

Fig. 1.2 The map $\Phi_2 : \mathrm{SL}(2, \mathbb{C}) \to \mathcal{L}_+^\uparrow$, defined by $[\Phi_2(G)](x) := \sigma^{-1}(G\,\sigma(x)\,G^*)$, for every $G \in \mathrm{SL}(2, \mathbb{C})$ and $x \in \mathbb{R}^4$, is a double covering of $\mathcal{L}_+^\uparrow$

Proof For this purpose, let

$$G = \begin{pmatrix} \alpha & \beta \\ \gamma & \delta \end{pmatrix} \in \mathrm{SL}(2, \mathbb{C}) \; .$$

First, it follows that

$$\begin{aligned}
&[(\sigma^{-1} \circ L_G \circ \sigma)(x)] \cdot [(\sigma^{-1} \circ L_G \circ \sigma)(x)] \\
&= \det\big(\sigma((\sigma^{-1} \circ L_G \circ \sigma)(x))\big) = \det(L_G(\sigma(x))) \\
&= \det\big(G\, \sigma(x)\, G^*\big) = \det(\sigma(x)) = x \cdot x \; ,
\end{aligned}$$

for every $x \in \mathbb{R}^4$, and hence that $\sigma^{-1} \circ L_G \circ \sigma$ is a Lorentz transformation. Further, it follows from some calculation that

$$(\sigma^{-1} \circ L_G \circ \sigma)(x) = \Lambda_G \cdot x \; ,$$

for every $x \in \mathbb{R}^4$, where

$$\Lambda_G = \frac{1}{2} \begin{pmatrix}
|\alpha|^2 + |\beta|^2 + |\gamma|^2 + |\delta|^2 & \alpha^*\beta + \alpha\beta^* + \gamma^*\delta + \gamma\delta^* & i\,(\alpha^*\beta - \alpha\beta^* + \gamma^*\delta - \gamma\delta^*) & |\alpha|^2 - |\beta|^2 + |\gamma|^2 - |\delta|^2 \\
\alpha^*\gamma + \alpha\gamma^* + \beta^*\delta + \beta\delta^* & \alpha^*\delta + \alpha\delta^* + \beta^*\gamma + \beta\gamma^* & i\,(\alpha^*\delta - \alpha\delta^* - \beta^*\gamma + \beta\gamma^*) & \alpha^*\gamma + \alpha\gamma^* - \beta^*\delta - \beta\delta^* \\
i\,(-\alpha^*\gamma + \alpha\gamma^* - \beta^*\delta + \beta\delta^*) & i\,(-\alpha^*\delta + \alpha\delta^* - \beta^*\gamma + \beta\gamma^*) & \alpha^*\delta + \alpha\delta^* - \beta^*\gamma - \beta\gamma^* & i\,(-\alpha^*\gamma + \alpha\gamma^* + \beta^*\delta - \beta\delta^*) \\
|\alpha|^2 + |\beta|^2 - |\gamma|^2 - |\delta|^2 & \alpha^*\beta + \alpha\beta^* - \gamma^*\delta - \gamma\delta^* & i\,(\alpha^*\beta - \alpha\beta^* - \gamma^*\delta + \gamma\delta^*) & |\alpha|^2 - |\beta|^2 - |\gamma|^2 + |\delta|^2
\end{pmatrix} .$$

Since $(\Lambda_G)_{00} \geqslant 0$, Λ_G preserves time orientation. In a second step, we notice that for $G_1, G_2 \in \mathrm{SL}(2, \mathbb{C})$, it follows that

$$\begin{aligned}
L_{G_2G_1}(A) &= (G_2G_1)A\,(G_2G_1)^* = G_2G_1A\,G_1^*G_2^* = G_2(G_1A\,G_1^*)G_2^* \\
&= G_2L_{G_1}(A)G_2^* = L_{G_2}(L_{G_1}(A)) = (L_{G_2} \circ L_{G_1})(A) \; ,
\end{aligned}$$

for all $A \in \mathrm{H}(2, \mathbb{C})$ and hence that

$$L_{G_2G_1} = L_{G_2} \circ L_{G_1} \; .$$

This also implies that

$$\begin{aligned}
\Lambda_{G_2G_1} &= \sigma^{-1} \circ L_{G_2G_1} \circ \sigma = \sigma^{-1} \circ L_{G_2} \circ \sigma \circ \sigma^{-1} \circ L_{G_1} \circ \sigma \\
&= \Lambda_{G_2} \cdot \Lambda_{G_1} \; .
\end{aligned} \tag{1.2}$$

In particular, we have the following decomposition of G. If $\gamma \neq 0$,

$$G = \begin{pmatrix} 1 & \alpha/\gamma \\ 0 & 1 \end{pmatrix} \cdot \begin{pmatrix} i/\gamma & 0 \\ 0 & -i\gamma \end{pmatrix} \cdot \begin{pmatrix} 0 & i \\ i & 0 \end{pmatrix} \cdot \begin{pmatrix} 1 & \delta/\gamma \\ 0 & 1 \end{pmatrix} \tag{1.3}$$

If $\gamma = 0$,

$$G = \begin{pmatrix} \alpha & 0 \\ 0 & \delta \end{pmatrix} \cdot \begin{pmatrix} 1 & \beta/\alpha \\ 0 & 1 \end{pmatrix} \; . \tag{1.4}$$

As a consequence, we consider 3 cases,

(1) $\alpha = \delta = 1, \gamma = 0$,
(2) $\beta = \gamma = 0$,
(3) $\alpha = \delta = 0, \beta = \gamma = i$.

In the first case,

$$\Lambda_G = \frac{1}{2}\begin{pmatrix} |\beta|^2 + 2 & \beta + \beta^* & i\,(\beta - \beta^*) & -|\beta|^2 \\ \beta + \beta^* & 2 & 0 & -(\beta + \beta^*) \\ i\,(\beta - \beta^*) & 0 & 2 & i\,(\beta^* - \beta) \\ |\beta|^2 & \beta + \beta^* & i\,(\beta - \beta^*) & 2 - |\beta|^2 \end{pmatrix}.$$

By addition of the negative of the last row to the first row, we arrive at

$$\frac{1}{2}\begin{pmatrix} 2 & 0 & 0 & -2 \\ \beta + \beta^* & 2 & 0 & -(\beta + \beta^*) \\ i\,(\beta - \beta^*) & 0 & 2 & i\,(\beta^* - \beta) \\ |\beta|^2 & \beta + \beta^* & i\,(\beta - \beta^*) & 2 - |\beta|^2 \end{pmatrix}.$$

By addition of the last column to the first, we arrive at

$$\frac{1}{2}\begin{pmatrix} 0 & 0 & 0 & -2 \\ 0 & 2 & 0 & -(\beta + \beta^*) \\ 0 & 0 & 2 & i\,(\beta^* - \beta) \\ 2 & \beta + \beta^* & i\,(\beta - \beta^*) & 2 - |\beta|^2 \end{pmatrix}.$$

By exchanging the first row and the fourth row, in this way changing the sign of the determinant of the matrix, we arrive at the upper triangular matrix

$$\frac{1}{2}\begin{pmatrix} 2 & \beta + \beta^* & i\,(\beta - \beta^*) & 2 - |\beta|^2 \\ 0 & 2 & 0 & -(\beta + \beta^*) \\ 0 & 0 & 2 & i\,(\beta^* - \beta) \\ 0 & 0 & 0 & -2 \end{pmatrix}.$$

The determinant of the latter matrix is given by -1. Hence in this case, we have that $\det(\Lambda_G) = 1$. *In the second case,*

$$\Lambda_G = \frac{1}{2}\begin{pmatrix} |\alpha|^2 + |\delta|^2 & 0 & 0 & |\alpha|^2 - |\delta|^2 \\ 0 & \alpha^*\delta + \alpha\delta^* & i\,(\alpha^*\delta - \alpha\delta^*) & 0 \\ 0 & -i\,(\alpha^*\delta - \alpha\delta^*) & \alpha^*\delta + \alpha\delta^* & 0 \\ |\alpha|^2 - |\delta|^2 & 0 & 0 & |\alpha|^2 + |\delta|^2 \end{pmatrix}.$$

The determinant of the latter matrix is given by

$$
\begin{aligned}
&\frac{1}{2^4}\left[(|\alpha|^2+|\delta|^2)^2-(|\alpha|^2-|\delta|^2)^2\right]\cdot\left[(\alpha^*\delta+\alpha\delta^*)^2-(\alpha^*\delta-\alpha\delta^*)^2\right] \\
&=\frac{1}{2^4}\,4\,|\alpha|^2\,|\delta|^2\,4\,\alpha^*\delta\,\alpha\,\delta^*=|\alpha|^4\,|\delta|^4=|\alpha\cdot\delta|^4=1\ .
\end{aligned}
$$

Hence also in this case, $\det(\Lambda_G)=1$. For subsequent use, we consider two special cases. The case,

$$
\alpha=e^{s/2}\ ,\ \delta=e^{-s/2}
$$

for some $s\in\mathbb{R}$, leads to the matrix

$$
\Lambda_1(s)=\begin{pmatrix}\cosh(s) & 0 & 0 & \sinh(s)\\ 0 & 1 & 0 & 0\\ 0 & 0 & 1 & 0\\ \sinh(s) & 0 & 0 & \cosh(s)\end{pmatrix}.
$$

The case,

$$
\alpha=e^{-i\varphi/2}\ ,\ \delta=e^{i\varphi/2}\ ,
$$

for some $\varphi\in\mathbb{R}$, leads to the matrix

$$
\Lambda_2(\varphi)=\begin{pmatrix}1 & 0 & 0 & 0\\ 0 & \cos(\varphi) & -\sin(\varphi) & 0\\ 0 & \sin(\varphi) & \cos(\varphi) & 0\\ 0 & 0 & 0 & 1\end{pmatrix}.
$$

In the third case,

$$
\Lambda_G=\frac{1}{2}\begin{pmatrix}2 & 0 & 0 & 0\\ 0 & 2 & 0 & 0\\ 0 & 0 & -2 & 0\\ 0 & 0 & 0 & -2\end{pmatrix}=\begin{pmatrix}1 & 0 & 0 & 0\\ 0 & 1 & 0 & 0\\ 0 & 0 & -1 & 0\\ 0 & 0 & 0 & -1\end{pmatrix}.
$$

The determinant of the latter matrix is given by 1. Hence also in this 3rd case, $\det(\Lambda_G)=1$. As a consequence, it follows from (1.2), (1.3) and (1.4) that $\det(\Lambda_G)=1$ for all $G\in\mathrm{SL}(2,\mathbb{C})$ and that $\sigma^{-1}\circ L_G\circ\sigma$ is a restricted Lorentz transformation. We note that the Λ_G corresponding to the case

$$
\alpha=\gamma=\frac{1}{2}(1+i)\ ,\ \beta=\frac{1}{2}(1-i)\ ,\ \delta=\frac{1}{2}(-1+i)=-\beta
$$

is given by

$$
\Lambda_3=\begin{pmatrix}1 & 0 & 0 & 0\\ 0 & 0 & 0 & 1\\ 0 & 1 & 0 & 0\\ 0 & 0 & 1 & 0\end{pmatrix}.
$$

Hence, we conclude from (1.2), Proposition 4.8 of Part I, a decomposition of rotations using proper Euler angles, since

$$\Lambda_3 \cdot \Lambda_2(\varphi) \cdot \Lambda_3^2 = \begin{pmatrix} 1 & 0 & 0 & 0 \\ 0 & 1 & 0 & 0 \\ 0 & 0 & \cos(\varphi) & -\sin(\varphi) \\ 0 & 0 & \sin(\varphi) & \cos(\varphi) \end{pmatrix} ,$$

for every $\varphi \in \mathbb{R}$, and from Theorem 5.4 of Part I, a decomposition of the members of $\mathcal{L}_+^\uparrow$, that by

$$\Phi_2(G) := \sigma^{-1} \circ L_G \circ \sigma$$

for every $G \in \mathrm{SL}(2, \mathbb{C})$, there is given an epimorphism $\Phi_2 : \mathrm{SL}(2, \mathbb{C}) \to \mathcal{L}_+^\uparrow$. Further, if $G_1, G_2 \in \mathrm{SL}(2, \mathbb{C})$ are such that $\Phi_2(G_1) = \Phi_2(G_2)$, then $\Phi_2(G_1 G_2^{-1})$ is given by the unit matrix. If $\alpha, \beta, \gamma, \delta \in \mathbb{C}$ are such that

$$G_1 G_2^{-1} = \begin{pmatrix} \alpha & \beta \\ \gamma & \delta \end{pmatrix} ,$$

The latter leads to

$$|\alpha|^2 + |\beta|^2 = 1 \ , \ |\alpha|^2 - |\beta|^2 = 1 \ ,$$
$$|\delta|^2 + |\gamma|^2 = 1 \ , \ |\delta|^2 - |\gamma|^2 = 1$$

which implies that

$$\alpha, \delta \in S^1 \ , \ \beta = \gamma = 0 \ .$$

Hence it follows from Case (2) above that

$$\alpha^* \delta + \alpha \, \delta^* = 2 \ , \ \alpha^* \delta - \alpha \, \delta^* = 0 \ ,$$

which implies that

$$\alpha^* \delta = 1 \ .$$

From the latter, it follows that $\delta = \alpha\alpha^*\delta = \alpha$. Finally, since $\alpha\delta = 1$, we conclude that $\alpha = \pm 1$. Hence it follows that $G_1 G_2^{-1} = \pm E$. The latter implies that $G_2 = \pm G_1$. Finally, obviously, $\Phi_2(G) = \Phi_2(-G)$ for every $G \in \mathrm{SL}(2, \mathbb{C})$. □

From the proofs of Theorem 4.13 of Part I and Theorem 1.2, we obtain the following corollaries (Fig. 1.3).

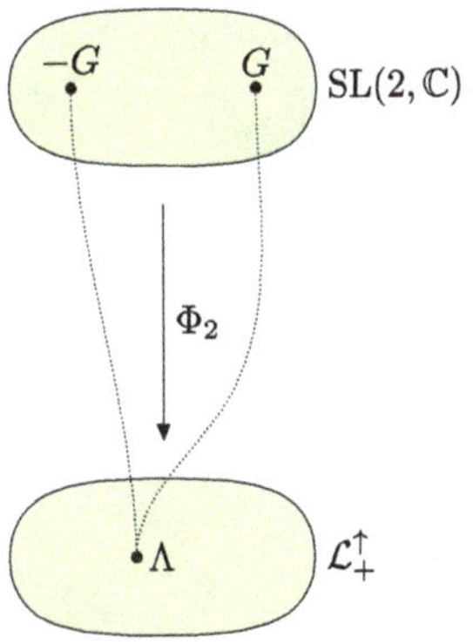

Fig. 1.3 Depiction of the action of Φ_2

Corollary 1.3 (Particular images of Φ_2) *For $s, \psi \in \mathbb{R}$, we have that*

$$\Phi_2\left(\begin{pmatrix} e^{s/2} & 0 \\ 0 & e^{-s/2} \end{pmatrix}\right) = \begin{pmatrix} \cosh(s) & 0 & 0 & \sinh(s) \\ 0 & 1 & 0 & 0 \\ 0 & 0 & 1 & 0 \\ \sinh(s) & 0 & 0 & \cosh(s) \end{pmatrix},$$

$$\Phi_2\left(\begin{pmatrix} e^{i\psi/2} & 0 \\ 0 & e^{-i\psi/2} \end{pmatrix}\right) = \begin{pmatrix} 1 & 0 & 0 & 0 \\ 0 & \cos(\psi) & \sin(\psi) & 0 \\ 0 & -\sin(\psi) & \cos(\psi) & 0 \\ 0 & 0 & 0 & 1 \end{pmatrix}, \tag{1.5}$$

$$\Phi_2\left(\frac{1}{\sqrt{2}}\begin{pmatrix} e^{i\pi/4} & e^{-i\pi/4} \\ e^{i\pi/4} & -e^{-i\pi/4} \end{pmatrix}\right) = \begin{pmatrix} 1 & 0 & 0 & 0 \\ 0 & 0 & 0 & 1 \\ 0 & 1 & 0 & 0 \\ 0 & 0 & 1 & 0 \end{pmatrix}$$

and, for

$$G = \begin{pmatrix} \alpha & \beta \\ \gamma & \delta \end{pmatrix} \in \mathrm{SL}(2, \mathbb{C}),$$

we have that

$$\Phi_2(G) \tag{1.6}$$
$$= \begin{pmatrix} \frac{1}{2}(|\alpha|^2 + |\beta|^2 + |\gamma|^2 + |\delta|^2) & \mathrm{Re}(\alpha\beta^* + \gamma\delta^*) & \mathrm{Im}(\alpha\beta^* + \gamma\delta^*) & \frac{1}{2}(|\alpha|^2 - |\beta|^2 + |\gamma|^2 - |\delta|^2) \\ \mathrm{Re}(\alpha\gamma^* + \beta\delta^*) & \mathrm{Re}(\alpha\delta^* + \beta\gamma^*) & \mathrm{Im}(\alpha\delta^* - \beta\gamma^*) & \mathrm{Re}(\alpha\gamma^* - \beta\delta^*) \\ -\mathrm{Im}(\alpha\gamma^* + \beta\delta^*) & -\mathrm{Im}(\alpha\delta^* + \beta\gamma^*) & \mathrm{Re}(\alpha\delta^* - \beta\gamma^*) & -\mathrm{Im}(\alpha\gamma^* - \beta\delta^*) \\ \frac{1}{2}(|\alpha|^2 + |\beta|^2 - |\gamma|^2 - |\delta|^2) & \mathrm{Re}(\alpha\beta^* - \gamma\delta^*) & \mathrm{Im}(\alpha\beta^* - \gamma\delta^*) & \frac{1}{2}(|\alpha|^2 - |\beta|^2 - |\gamma|^2 + |\delta|^2) \end{pmatrix},$$

In particular, if $\alpha, \beta \in \mathbb{C}$ *are such that* $|\alpha|^2 + |\beta|^2 = 1$,

$$G := \begin{pmatrix} \alpha & \beta \\ -\beta^* & \alpha^* \end{pmatrix} (\in \mathrm{SU}(2)) ,$$

then

$$\Phi_2(G) = \begin{pmatrix} 1 & 0 & 0 & 0 \\ 0 & \mathrm{Re}(\alpha^2 - \beta^2) & \mathrm{Im}(\alpha^2 + \beta^2) & -2\,\mathrm{Re}(\alpha\beta) \\ 0 & -\mathrm{Im}(\alpha^2 - \beta^2) & \mathrm{Re}(\alpha^2 + \beta^2) & 2\,\mathrm{Im}(\alpha\beta) \\ 0 & 2\,\mathrm{Re}(\alpha\beta^*) & 2\,\mathrm{Im}(\alpha\beta^*) & |\alpha|^2 - |\beta|^2 \end{pmatrix}$$

$$= \begin{pmatrix} 1 & 0 & 0 & 0 \\ 0 & [\Phi_1(G)]_{11} & [\Phi_1(G)]_{12} & [\Phi_1(G)]_{13} \\ 0 & [\Phi_1(G)]_{21} & [\Phi_1(G)]_{22} & [\Phi_1(G)]_{23} \\ 0 & [\Phi_1(G)]_{31} & [\Phi_1(G)]_{32} & [\Phi_1(G)]_{33} \end{pmatrix} , \tag{1.7}$$

where $\Phi_1 : \mathrm{SU}(2) \to \mathrm{SO}(3)$ *is the double covering from Theorem 4.13 of Part I.*

As a consequence, by help of Corollary 4.14 from Part I, specifying particular images of the double covering Φ_1 of SO(3), it follows that

$$\Phi_2\left(\begin{pmatrix} \cos(\varphi/2) & \sin(\varphi/2) \\ -\sin(\varphi/2) & \cos(\varphi/2) \end{pmatrix}\right) = \begin{pmatrix} 1 & 0 & 0 & 0 \\ 0 & \cos(\varphi) & 0 & -\sin(\varphi) \\ 0 & 0 & 1 & 0 \\ 0 & \sin(\varphi) & 0 & \cos(\varphi) \end{pmatrix} ,$$

$$\Phi_2\left(\begin{pmatrix} \cos(\theta/2) & i\sin(\theta/2) \\ i\sin(\theta/2) & \cos(\theta/2) \end{pmatrix}\right) = \begin{pmatrix} 1 & 0 & 0 & 0 \\ 0 & 1 & 0 & 0 \\ 0 & 0 & \cos(\theta) & \sin(\theta) \\ 0 & 0 & -\sin(\theta) & \cos(\theta) \end{pmatrix} , \tag{1.8}$$

for all $\varphi, \theta \in \mathbb{R}$.

Corollary 1.4 (Continuity of Φ_2) *Let* $G \in \mathrm{SL}(2, \mathbb{C})$. *If* $G_1, G_2, \ldots$ *is a sequence in* $\mathrm{SL}(2, \mathbb{C})$ *that converges component-wise to* $G \in \mathrm{SL}(2, \mathbb{C})$, *then the corresponding sequence in* $\mathcal{L}_+^\uparrow$,

$$\Phi_2(G_1), \Phi_2(G_2), \ldots ,$$

converges component-wise to $\Phi_2(G)$.

Proof The statement is a simple consequence of the explicit representation (1.6) of Φ_2. □

Exercise 1.1 (i) Show that

$$U \cdot \begin{pmatrix} \cosh(s) & 0 & 0 & \sinh(s) \\ 0 & 1 & 0 & 0 \\ 0 & 0 & 1 & 0 \\ \sinh(s) & 0 & 0 & \cosh(s) \end{pmatrix} U^{-1} = \begin{pmatrix} \cosh(s) & \sinh(s) & 0 & 0 \\ \sinh(s) & \cosh(s) & 0 & 0 \\ 0 & 0 & 1 & 0 \\ 0 & 0 & 0 & 1 \end{pmatrix},$$

$$U \cdot \begin{pmatrix} \cosh(s) & \sinh(s) & 0 & 0 \\ \sinh(s) & \cosh(s) & 0 & 0 \\ 0 & 0 & 1 & 0 \\ 0 & 0 & 0 & 1 \end{pmatrix} U^{-1} = \begin{pmatrix} \cosh(s) & 0 & \sinh(s) & 0 \\ 0 & 1 & 0 & 0 \\ \sinh(s) & 0 & \cosh(s) & 0 \\ 0 & 0 & 0 & 1 \end{pmatrix},$$

for every $s \in \mathbb{R}$, where

$$U = \begin{pmatrix} 1 & 0 & 0 & 0 \\ 0 & 0 & 0 & 1 \\ 0 & 1 & 0 & 0 \\ 0 & 0 & 1 & 0 \end{pmatrix} \in \mathcal{L}_+^\uparrow .$$

(ii) With he help of (1.5), show that

$$\Phi_2\left(\begin{pmatrix} \cosh(s/2) & \sinh(s/2) \\ \sinh(s/2) & \cosh(s/2) \end{pmatrix}\right) = \begin{pmatrix} \cosh(s) & \sinh(s) & 0 & 0 \\ \sinh(s) & \cosh(s) & 0 & 0 \\ 0 & 0 & 1 & 0 \\ 0 & 0 & 0 & 1 \end{pmatrix},$$

$$\Phi_2\left(\begin{pmatrix} \cosh(s/2) & -i\sinh(s/2) \\ i\sinh(s/2) & \cosh(s/2) \end{pmatrix}\right) = \begin{pmatrix} \cosh(s) & 0 & \sinh(s) & 0 \\ 0 & 1 & 0 & 0 \\ \sinh(s) & 0 & \cosh(s) & 0 \\ 0 & 0 & 0 & 1 \end{pmatrix},$$

for every $s \in \mathbb{R}$.

Further, from Exercise 1.1, it follows, for $s, \theta, \varphi, \psi \in \mathbb{R}$, that

$$\Phi_2\left(\begin{pmatrix} \cosh(s/2) & \sinh(s/2) \\ \sinh(s/2) & \cosh(s/2) \end{pmatrix}\right) = \begin{pmatrix} \cosh(s) & \sinh(s) & 0 & 0 \\ \sinh(s) & \cosh(s) & 0 & 0 \\ 0 & 0 & 1 & 0 \\ 0 & 0 & 0 & 1 \end{pmatrix},$$

$$\Phi_2\left(\begin{pmatrix} \cosh(s/2) & -i\sinh(s/2) \\ i\sinh(s/2) & \cosh(s/2) \end{pmatrix}\right) = \begin{pmatrix} \cosh(s) & 0 & \sinh(s) & 0 \\ 0 & 1 & 0 & 0 \\ \sinh(s) & 0 & \cosh(s) & 0 \\ 0 & 0 & 0 & 1 \end{pmatrix}, \tag{1.9}$$

Further, for $x \in \mathbb{R}^4$, we have that

$$\sigma(\Phi_2(G)(x)) = G \cdot \sigma(x) \cdot G^* \ . \tag{1.10}$$

Reminding the reader that we identify the elements of $L(\mathbb{R}^4, \mathbb{R}^4)$, with their corresponding representation matrices, with respect to $e_0, \dots, e_3$, where $e_0 := {}^t(1, 0, 0, 0)$, $e_1 := {}^t(0, 1, 0, 0)$, $e_2 := {}^t(0, 0, 1, 0)$ and $e_3 := {}^t(0, 0, 0, 1)$, it follows from the Definition (1.1) of σ and (1.10) that

$$\sum_{k=0}^{3} (\Phi_2(G) \cdot x)_k \, \sigma_k = G \cdot \left(\sum_{l=0}^{3} x_l \, \sigma_l \right) \cdot G^* \ .$$

From the latter, we infer that

$$\begin{aligned} \sum_{k=0}^{3} \sum_{l=0}^{3} [\Phi_2(G)]_{kl} \cdot x_l \, \sigma_k &= \sum_{l=0}^{3} \sum_{k=0}^{3} [\Phi_2(G)]_{kl} \cdot x_l \, \sigma_k \\ &= G \cdot \left(\sum_{l=0}^{3} x_l \, \sigma_l \right) \cdot G^* = \sum_{l=0}^{3} x_l \, G \cdot \sigma_l \cdot G^* \ , \end{aligned}$$

for ever $x \in \mathbb{R}^4$ and hence that

$$\boxed{\sum_{k=0}^{3} [\Phi_2(G)]_{kl} \, \sigma_k = G \cdot \sigma_l \cdot G^* \ ,} \tag{1.11}$$

for all $l \in \{0, 1, 2, 3\}$.

Exercise 1.2 Show for every $G \in \mathrm{SL}(2, \mathbb{C})$ that

$$\boxed{\Phi_2(G^*) = [\Phi_2(G)]^t \ ,} \tag{1.12}$$

and as consequence that, in addition to (1.11), for every $G \in \mathrm{SL}(2, \mathbb{C})$, we have that

$$\boxed{\sum_{k=0}^{3} [\Phi_2(G)]_{lk} \, \sigma_k = G^* \cdot \sigma_l \cdot G \ ,} \tag{1.13}$$

for all $l \in \{0, 1, 2, 3\}$.

Exercise 1.3 Show the following.

(i) By $\hbar_{cc}(G) := G_{cc} := (G^t)^*$, for every $G \in \mathrm{SL}(2, \mathbb{C})$, there is defined involutory group automorphism $\hbar_{cc} : \mathrm{SL}(2, \mathbb{C}) \to \mathrm{SL}(2, \mathbb{C})$ such that $\hbar_{cc}(-G) = -\hbar_{cc}(G)$, for every $G \in \mathrm{SL}(2, \mathbb{C})$, and $\hbar_{cc}(\mathrm{SU}(2, \mathbb{C})) \subset \mathrm{SU}(2, \mathbb{C})$. Further, $\hbar_{cc}$ is component-wise continuous, i.e., if $G_1, G_2, \dots$ is a sequence in $\mathrm{SL}(2, \mathbb{C})$ that converges component-wise to $G \in \mathrm{SL}(2, \mathbb{C})$, then the corresponding sequence $\hbar_{cc}(G_1), \hbar_{cc}(G_2), \ldots$ converges component-wise to $\hbar_{cc}(G)$.

(ii) By $\hbar_{*^{-1}}(G) := (G^*)^{-1}$, for every $G \in \mathrm{SL}(2, \mathbb{C})$, there is defined group automorphism $\hbar_{*^{-1}} : \mathrm{SL}(2, \mathbb{C}) \to \mathrm{SL}(2, \mathbb{C})$, such that and $\hbar_{*^{-1}}(-G) = -\hbar_{*^{-1}}(G)$, for every $G \in \mathrm{SL}(2, \mathbb{C})$, and that leaves the elements of $\mathrm{SU}(2, \mathbb{C})$ fixed. Further $\hbar_{*^{-1}}$ is component-wise continuous, i.e., if $G_1, G_2, \ldots$ is a sequence in $\mathrm{SL}(2, \mathbb{C})$ that converges component-wise to $G \in \mathrm{SL}(2, \mathbb{C})$, then the corresponding sequence $\hbar_{*^{-1}}(G_1), \hbar_{*^{-1}}(G_2), \ldots$ converges component-wise to $\hbar_{*^{-1}}(G)$. Finally, show that $(G^{-1})^* = (G^*)^{-1}$, for every $G \in \mathrm{SL}(2, \mathbb{C})$.

1.2 Further Properties of the Lorentz Group

We note that from (1.5), (1.8), (1.9), and the later Exercise 1.4, it follows that

$$\Phi_2(\exp((s/2)\,\sigma_1)) = \begin{pmatrix} \cosh(s) & \sinh(s) & 0 & 0 \\ \sinh(s) & \cosh(s) & 0 & 0 \\ 0 & 0 & 1 & 0 \\ 0 & 0 & 0 & 1 \end{pmatrix} , \quad \Phi_2(\exp((s/2)\,\sigma_2)) = \begin{pmatrix} \cosh(s) & 0 & \sinh(s) & 0 \\ 0 & 1 & 0 & 0 \\ \sinh(s) & 0 & \cosh(s) & 0 \\ 0 & 0 & 0 & 1 \end{pmatrix} ,$$

$$\Phi_2(\exp((s/2)\,\sigma_3)) = \begin{pmatrix} \cosh(s) & 0 & 0 & \sinh(s) \\ 0 & 1 & 0 & 0 \\ 0 & 0 & 1 & 0 \\ \sinh(s) & 0 & 0 & \cosh(s) \end{pmatrix} , \quad \Phi_2(\exp(i\,(s/2)\,\sigma_1)) = \begin{pmatrix} 1 & 0 & 0 & 0 \\ 0 & 1 & 0 & 0 \\ 0 & 0 & \cos(s) & \sin(s) \\ 0 & 0 & -\sin(s) & \cos(s) \end{pmatrix} ,$$

$$\Phi_2(\exp(i\,(s/2)\,\sigma_2)) = \begin{pmatrix} 1 & 0 & 0 & 0 \\ 0 & \cos(s) & 0 & -\sin(s) \\ 0 & 0 & 1 & 0 \\ 0 & \sin(s) & 0 & \cos(s) \end{pmatrix} , \quad \Phi_2(\exp(i\,(s/2)\,\sigma_3)) = \begin{pmatrix} 1 & 0 & 0 & 0 \\ 0 & \cos(s) & \sin(s) & 0 \\ 0 & -\sin(s) & \cos(s) & 0 \\ 0 & 0 & 0 & 1 \end{pmatrix} ,$$

for every $s \in \mathbb{R}$.

In the following, we are going to show that

$$\boxed{\exp(T_E\mathcal{L}) \subset \mathcal{L}_+^{\uparrow}\ .}$$

For the proof, we note for $A \in T_{E_{2\times2}}\mathrm{SL}(2, \mathbb{C})$ that, since $\Phi_2 : \mathrm{SL}(2, \mathbb{C}) \to \mathcal{L}$ is a group homomorphism,

$$\begin{aligned} \Phi_2(\exp((s_1 + s_2)A)) &= \Phi_2(\exp(s_1 A + s_2 A)) = \Phi_2(\exp(s_1 A)\exp(s_2 A)) \\ &= \Phi_2(\exp(s_1 A))\,\Phi_2(\exp(s_2 A))\ , \\ \Phi_2(\exp(0A)) &= \Phi_2(E_{2\times2}) = E_{4\times4}\ , \end{aligned}$$

for all $s_1, s_2 \in \mathbb{R}$, where $E_{2\times2}$ denotes the unit matrix in $\mathrm{M}(2, \mathbb{C})$ and $E_{4\times4}$ denotes the unit matrix in $\mathrm{M}(4, \mathbb{R})$. Hence, by

$$T(s) := \Phi_2(\exp(sA)) ,$$

for every $s \in \mathbb{R}$, there is given a one-parameter subgroup of $\mathcal{L}$. Further, it follows from (1.6) that T is differentiable. If $\alpha, \beta, \gamma, \delta \in \mathbb{C}$ are such that

$$A = \begin{pmatrix} \alpha & \beta \\ \gamma & \delta \end{pmatrix} ,$$

note that since $\mathrm{Tr}(A) = 0$, we have that $\alpha + \delta = 0$, it follows after some calculation that

$$T'(0) = \begin{pmatrix} 0 & \beta_1 + \gamma_1 & -\beta_2 + \gamma_2 & 2\alpha_1 \\ \beta_1 + \gamma_1 & 0 & 2\alpha_2 & -(\beta_1 - \gamma_1) \\ -\beta_2 + \gamma_2 & -2\alpha_2 & 0 & \beta_2 + \gamma_2 \\ 2\alpha_1 & \beta_1 - \gamma_1 & -(\beta_2 + \gamma_2) & 0 \end{pmatrix} \in T_{E_{4\times4}}\mathcal{L} ,$$

where $\alpha_1, \beta_1, \gamma_1$ and δ_1 are the real parts of α, β, γ and δ, respectively, and $\alpha_2, \beta_2, \gamma_2$ and δ_2 are the imaginary parts of α, β, γ and δ, respectively. Hence, it follows from Theorem 4.5.1 (ii) of [4] and Theorem 3.8 (iii) of Part I, providing properties of the exponential map, that the one-parameter groups T and

$$(\mathbb{R} \to \mathrm{M}(4, \mathbb{R}), s \mapsto \exp(sT'(0)))$$

coincide. In particular, we have that

$$\exp(T'(0)) = \Phi_2(\exp(A)) \in \mathcal{L}_+^\uparrow .$$

Further, if $a, b \in \mathbb{R}^3$ and

$$\alpha := \frac{1}{2}(a_3 - ib_3) , \quad \beta := \frac{1}{2}[a_1 - ia_2 - i(b_1 - ib_2)] ,$$
$$\gamma := \frac{1}{2}[a_1 + ia_2 - i(b_1 + ib_2)] , \quad \delta := -\alpha ,$$

then

$$A := \begin{pmatrix} \alpha & \beta \\ \gamma & \delta \end{pmatrix} \in T_{E_{2\times2}}\mathrm{SL}(2, \mathbb{C})$$

and

$$T'(0) = \begin{pmatrix} 0 & a_1 & a_2 & a_3 \\ a_1 & 0 & -b_3 & b_2 \\ a_2 & b_3 & 0 & -b_1 \\ a_3 & -b_2 & b_1 & 0 \end{pmatrix} \in T_{E_{4\times4}}\mathcal{L}$$

as well as

$$\exp\left(\begin{pmatrix} 0 & a_1 & a_2 & a_3 \\ a_1 & 0 & -b_3 & b_2 \\ a_2 & b_3 & 0 & -b_1 \\ a_3 & -b_2 & b_1 & 0 \end{pmatrix}\right) = \Phi_2(\exp(A)) \in \mathcal{L}_+^{\uparrow} .$$

1.3 Weyl Spinors

For every homomorphism $\hbar : \mathrm{SL}(2, \mathbb{C}) \to \mathrm{SL}(2, \mathbb{C})$ and $G \in \mathrm{SL}(2, \mathbb{C})$, we define

$$D^{\hbar}(G) := (\mathbb{C}^2 \to \mathbb{C}^2, \psi \mapsto \hbar(G) \cdot \psi) .$$

From the linearity of matrix multiplication, it follows that $D^{\hbar}(G)$ is linear.

Further,

$$D^{\hbar} := \begin{pmatrix} \mathrm{SL}(2, \mathbb{C}) \to & \mathrm{L}(\mathbb{C}^2) \\ G & \mapsto D^{\hbar}(G) \end{pmatrix} ,$$

is a representation of $\mathrm{SL}(2, \mathbb{C})$,

since for $G_1, G_2 \in \mathrm{SL}(2, \mathbb{C})$, using the associativity of matrix multiplication, we have that

$$\begin{aligned} & D^{\hbar}(G_1 G_2)\psi = \hbar(G_1 \cdot G_2) \cdot \psi = [\hbar(G_1) \cdot \hbar(G_2)] \cdot \psi \\ & = \hbar(G_1) \cdot (\hbar(G_2) \cdot \psi) = D^{\hbar}(G_1) D^{\hbar}(G_2)\psi , \\ & D^{\hbar}(E)\psi = \hbar(E)\psi = E\psi = \psi , \end{aligned}$$

for every $\psi \in \mathbb{C}^2$, where E denotes the 2×2 unit matrix, and hence that

$$D^{\hbar}(G_1 G_2) = D^{\hbar}(G_1) \circ D^{\hbar}(G_2) , \; D^{\hbar}(E) = \mathrm{id}_{\mathbb{C}^2} .$$

In particular, since the representation $D^{\frac{1}{2}} : \mathrm{SU}(2) \to \mathrm{L}(\mathbb{C}^2)$ is irreducible, we note that

if $\mathrm{Ran}(\hbar) \supset \mathrm{SU}(2)$, the representation $D^{\hbar}$ is irreducible, i.e., there is no non-trivial proper subspace of $\mathbb{C}^2$ that is left invariant by every $D^{\hbar}(G)$, where G runs through the elements of $\mathrm{SL}(2, \mathbb{C})$.

From Exercise 1.3, it follows that by

$$\hbar_R(G) := G \tag{1.14}$$

and

$$\hbar_L(G) := (G^*)^{-1} = \begin{pmatrix} \delta^* & -\gamma^* \\ -\beta^* & \alpha^* \end{pmatrix} , \tag{1.15}$$

for every

$$G = \begin{pmatrix} \alpha & \beta \\ \gamma & \delta \end{pmatrix} \in \mathrm{SL}(2, \mathbb{C}) ,$$

there are defined group automorphisms that leave the elements of $\mathrm{SU}(2, \mathbb{C})$ fixed.

The elements of the representation space $\mathbb{C}^2$ of $D^{\hbar_R}$ and $D^{\hbar_L}$ are $\mathrm{SL}(2, \mathbb{C})$ two-component ("Weyl"-) spinors.

We note that, from the representations (1.14) and (1.15), for $G \in \mathrm{SL}(2, \mathbb{C})$, it follows that

$$\|\hbar_R(G)\|_\infty = \|\hbar_L(G)\|_\infty = \|G\|_\infty ,$$

where we define

$$\|G\|_\infty := \max\{|G_{11}|, |G_{12}|, |G_{21}|, |G_{22}|\} ,$$

for every $G \in \mathrm{M}(2, \mathbb{C})$. In addition, by the same argument, it follows that $\hbar_R$ and $\hbar_L$ are continuous, i.e., if $G_1, G_2, \dots$ is a sequence in $\mathrm{SL}(2, \mathbb{C})$ that converges component-wise to $G \in \mathrm{SL}(2, \mathbb{C})$, then the sequences $\hbar_R(G_1), \hbar_R(G_2), \dots$ and $\hbar_L(G_1), \hbar_L(G_2), \dots$ converge component-wise to $\hbar_R(G)$ and $\hbar_L(G)$, respectively.

1.4 Weyl Representations of $\mathrm{SL}(2, \mathbb{C})$

In the following, let $a \geqslant 0$ and $\hbar \in \{\hbar_R, \hbar_L\}$. For $G \in \mathrm{SL}(2, \mathbb{C})$, we define $T_a^\hbar(G)$ by

$$\begin{aligned} T_a^\hbar(G) \begin{pmatrix} f_1 \\ f_2 \end{pmatrix} &:= \hbar(G) \cdot \begin{pmatrix} U_a(\Phi_2(G)) f_1 \\ U_a(\Phi_2(G)) f_2 \end{pmatrix} \\ &= \begin{pmatrix} [\hbar(G)]_{11} U_a(\Phi_2(G)) f_1 + [\hbar(G)]_{12} U_a(\Phi_2(G)) f_2 \\ [\hbar(G)]_{21} U_a(\Phi_2(G)) f_1 + [\hbar(G)]_{22} U_a(\Phi_2(G)) f_2 \end{pmatrix} , \end{aligned} \tag{1.16}$$

for every ${}^t(f_1, f_2) \in (L^2_{\mathbb{C}}(\mathbb{R}^3, \varphi_a))^2$, where $U_a : \mathcal{L}_+^\uparrow \to \mathrm{L}(L^2_{\mathbb{C}}(\mathbb{R}^3, \varphi_a), L^2_{\mathbb{C}}(\mathbb{R}^3, \varphi_a))$ is the strongly continuous unitary representation of $\mathcal{L}_+^\uparrow$ from Part I, see Theorem 5.7 in Part I. In particular, the additive, monotone and regular interval function φ_a is defined by

$$\varphi_a(I) := \int_{\mathbb{R}^3} \chi_I \cdot (v_a^0)^{-1}\, dv^3 \ ,$$

for every bounded interval I of $\mathbb{R}^3$, where $v_a^0 : \mathbb{R}^3 \to \mathbb{R}$ is defined by

$$v_a^0(v) := (|v|^2 + a^2)^{1/2} \ ,$$

for every $v \in \mathbb{R}^3$, and v^3 denotes the Lebesgue measure on $\mathbb{R}^3$, such that

- a subset $N \subset \mathbb{R}^3$ is a φ_a-zero set if and only if it is a v^3-zero set
- and that g is φ_a-integrable if and only if $(v_a^0)^{-1} g$ is v^3-integrable and that in this case we have that
$$\int_{\mathbb{R}^3} g\, d\varphi_a = \int_{\mathbb{R}^3} g \cdot (v_a^0)^{-1}\, dv^3 \ .$$

The map $T_a^{\hbar}(G)$ is obviously linear and satisfies

$$\begin{aligned}
&\langle T_a^{\hbar}(G)\,{}^t(f_1, f_2) | T_a^{\hbar}(G)\,{}^t(g_1, g_2)\rangle \\
&= \langle [\hbar(G)]_{11} U_a(\Phi_2(G)) f_1 + [\hbar(G)]_{12} U_a(\Phi_2(G)) f_2 | [\hbar(G)]_{11} U_a(\Phi_2(G)) g_1 + [\hbar(G)]_{12} U_a(\Phi_2(G)) g_2\rangle_2 \\
&\quad + \langle [\hbar(G)]_{21} U_a(\Phi_2(G)) f_1 + [\hbar(G)]_{22} U_a(\Phi_2(G)) f_2 | [\hbar(G)]_{21} U_a(\Phi_2(G)) g_1 + [\hbar(G)]_{22} U_a(\Phi_2(G)) g_2\rangle_2 \\
&= |[\hbar(G)]_{11}|^2 \langle f_1|g_1\rangle_2 + [\hbar(G)]_{11}^* [\hbar(G)]_{12} \langle f_1|g_2\rangle_2 + [\hbar(G)]_{12}^* [\hbar(G)]_{11} \langle f_2|g_1\rangle_2 + |[\hbar(G)]_{12}|^2 \langle f_2|g_2\rangle_2 \\
&\quad + |[\hbar(G)]_{21}|^2 \langle f_1|g_1\rangle_2 + [\hbar(G)]_{21}^* [\hbar(G)]_{22} \langle f_1|g_2\rangle_2 + [\hbar(G)]_{22}^* [\hbar(G)]_{21} \langle f_2|g_1\rangle_2 + |[\hbar(G)]_{22}|^2 \langle f_2|g_2\rangle_2 \\
&= (|[\hbar(G)]_{11}|^2 + |[\hbar(G)]_{21}|^2) \langle f_1|g_1\rangle_2 + ([\hbar(G)]_{11}^* [\hbar(G)]_{12} + [\hbar(G)]_{21}^* [\hbar(G)]_{22}) \langle f_1|g_2\rangle_2 \\
&\quad + ([\hbar(G)]_{11} [\hbar(G)]_{12}^* + [\hbar(G)]_{21} [\hbar(G)]_{22}^*) \langle f_2|g_1\rangle_2 + (|[\hbar(G)]_{12}|^2 + |[\hbar(G)]_{22}|^2) \langle f_2|g_2\rangle_2 \ ,
\end{aligned}$$

for all ${}^t(f_1, f_2), {}^t(g_1, g_2) \in (L^2_{\mathbb{C}}(\mathbb{R}^3, \varphi_a))^2$, where $\langle\,|\,\rangle_2$ denotes the scalar product for $L^2_{\mathbb{C}}(\mathbb{R}^3, \varphi_a)$, with induced norm $\|\ \|_2$ and the scalar product $\langle\,|\,\rangle$ on $(L^2_{\mathbb{C}}(\mathbb{R}^3, \varphi_a))^2$, with induced norm $\|\ \|$, is defined by

$$\langle {}^t(f_1, f_2) | {}^t(g_1, g_2)\rangle := \langle f_1|g_1\rangle_2 + \langle f_2|g_2\rangle_2 \ ,$$

for all ${}^t(f_1, f_2), {}^t(g_1, g_2) \in (L^2_{\mathbb{C}}(\mathbb{R}^3, \varphi_a))^2$. Hence if $G \in \mathrm{SU}(2)$, we have that $\hbar(G) = G \in \mathrm{SU}(2)$ and therefore that

$$\begin{aligned}
&\langle T_a^{\hbar}(G)\,{}^t(f_1, f_2) | T_a^{\hbar}(G)\,{}^t(g_1, g_2)\rangle \\
&= \langle f_1|g_1\rangle_2 + \langle f_2|g_2\rangle_2 = \langle {}^t(f_1, f_2) | {}^t(g_1, g_2)\rangle \ ,
\end{aligned}$$

for all ${}^t(f_1, f_2), {}^t(g_1, g_2) \in (L^2_{\mathbb{C}}(\mathbb{R}^3, \varphi_a))^2$, and $T_a^{\hbar}(G)$ preserves the scalar product. In the general case, we have that

$$
\begin{aligned}
&\|T_a^{\hbar}(G)\,{}^t(f_1, f_2)\|^2 = \langle T_a^{\hbar}(G)\,{}^t(f_1, f_2)|T_a^{\hbar}(G)\,{}^t(f_1, f_2)\rangle \\
&= (|[\hbar(G)]_{11}|^2 + |[\hbar(G)]_{21}|^2)\,\langle f_1|f_1\rangle_2 + ([\hbar(G)]_{11}^*[\hbar(G)]_{12} + [\hbar(G)]_{21}^*[\hbar(G)]_{22})\,\langle f_1|f_2\rangle_2 \\
&\quad + ([\hbar(G)]_{11}[\hbar(G)]_{12}^* + [\hbar(G)]_{21}[\hbar(G)]_{22}^*)\,\langle f_2|f_1\rangle_2 + (|[\hbar(G)]_{12}|^2 + |[\hbar(G)]_{22}|^2)\,\langle f_2|f_2\rangle_2 \\
&\leqslant 2\|G\|_\infty^2[\,\langle f_1|f_1\rangle_2 + \langle f_2|f_2\rangle_2 + |\,\langle f_1|f_2\rangle_2\,| + |\,\langle f_2|f_1\rangle_2\,|\,] \\
&\leqslant 2\|G\|_\infty^2[\,\|f_1\|_2^2 + \|f_2\|_2^2 + 2\,\|f_1\|_2\|f_2\|_2\,] \leqslant 4\|G\|_\infty^2(\|f_1\|_2^2 + \|f_2\|_2^2) = 4\|G\|_\infty^2\|\,{}^t(f_1, f_2)\|^2\,,
\end{aligned}
$$

for all ${}^t(f_1, f_2) \in (L^2_{\mathbb{C}}(\mathbb{R}^3, \varphi_a))^2$, and hence that by (1.16), there is defined a map

$$
T_a^{\hbar} : \mathrm{SL}(2, \mathbb{C}) \to \mathrm{L}((L^2_{\mathbb{C}}(\mathbb{R}^3, \varphi_a))^2, (L^2_{\mathbb{C}}(\mathbb{R}^3, \varphi_a))^2)\,.
$$

Further, for $G, H \in \mathrm{SL}(2, \mathbb{C})$ and ${}^t(f_1, f_2) \in (L^2_{\mathbb{C}}(\mathbb{R}^3, \varphi_a))^2$, we have that

$$
\begin{aligned}
&T_a^{\hbar}(G)\,T_a^{\hbar}(H)\begin{pmatrix} f_1 \\ f_2 \end{pmatrix} \\
&= T_a^{\hbar}(G)\begin{pmatrix} [\hbar(H)]_{11}U_a(\Phi_2(H))f_1 + [\hbar(H)]_{12}U_a(\Phi_2(H))f_2 \\ [\hbar(H)]_{21}U_a(\Phi_2(H))f_1 + [\hbar(H)]_{22}U_a(\Phi_2(H))f_2 \end{pmatrix} \\
&= \hbar(G)\cdot\begin{pmatrix} U_a(\Phi_2(G))[[\hbar(H)]_{11}U_a(\Phi_2(H))f_1 + [\hbar(H)]_{12}U_a(\Phi_2(H))f_2] \\ U_a(\Phi_2(G))[[\hbar(H)]_{21}U_a(\Phi_2(H))f_1 + [\hbar(H)]_{22}U_a(\Phi_2(H))f_2] \end{pmatrix} \\
&= \hbar(G)\cdot\begin{pmatrix} [\hbar(H)]_{11}U_a(\Phi_2(G\cdot H))f_1 + [\hbar(H)]_{12}U_a(\Phi_2(G\cdot H))f_2 \\ [\hbar(H)]_{21}U_a(\Phi_2(G\cdot H))f_1 + [\hbar(H)]_{22}U_a(\Phi_2(G\cdot H))f_2 \end{pmatrix} \\
&= \hbar(G)\cdot\hbar(H)\cdot\begin{pmatrix} U_a(\Phi_2(G\cdot H))f_1 \\ U_a(\Phi_2(G\cdot H))f_2 \end{pmatrix} = \hbar(G\cdot H)\cdot\begin{pmatrix} U_a(\Phi_2(G\cdot H))f_1 \\ U_a(\Phi_2(G\cdot H))f_2 \end{pmatrix} \\
&= T_a^{\hbar}(G\cdot H)\begin{pmatrix} f_1 \\ f_2 \end{pmatrix}
\end{aligned}
$$

and hence that

$$
T_a^{\hbar}(G\cdot H) = T_a^{\hbar}(G)\circ T_a^{\hbar}(H)\,.
$$

Since $T_a^{\hbar}(E)$ is given by the identity map on $(L^2_{\mathbb{C}}(\mathbb{R}^3, \varphi_a))^2$, it follows that $T_a^{\hbar}$ is a representation of $\mathrm{SL}(2, \mathbb{C})$. We note that this implies also that $T_a^{\hbar}(G)$ is unitary for every $G \in \mathrm{SU}(2)$. In this connection, also note Corollary 1.3. Further, if $G, H \in \mathrm{SL}(2, \mathbb{C})$ and ${}^t(f_1, f_2) \in (L^2_{\mathbb{C}}(\mathbb{R}^3, \varphi_a))^2$, we have that

$$T_a^{\hbar}(G)f - T_a^{\hbar}(H)f = \hbar(G) \cdot \begin{pmatrix} U_a(\Phi_2(G))f_1 \\ U_a(\Phi_2(G))f_2 \end{pmatrix} - \hbar(H) \cdot \begin{pmatrix} U_a(\Phi_2(H))f_1 \\ U_a(\Phi_2(H))f_2 \end{pmatrix}$$
$$= (\hbar(G) - \hbar(H)) \cdot \begin{pmatrix} U_a(\Phi_2(G))f_1 \\ U_a(\Phi_2(G))f_2 \end{pmatrix} + \hbar(H) \cdot \begin{pmatrix} [U_a(\Phi_2(G)) - U_a(\Phi_2(H))]f_1 \\ [U_a(\Phi_2(G)) - U_a(\Phi_2(H))]f_2 \end{pmatrix}$$
$$= \begin{pmatrix} ([\hbar(G)]_{11} - [\hbar(H)]_{11})U_a(\Phi_2(G))f_1 + ([\hbar(G)]_{12} - [\hbar(H)]_{12})U_a(\Phi_2(G))f_2 \\ ([\hbar(G)]_{21} - [\hbar(H)]_{21})U_a(\Phi_2(G))f_1 + ([\hbar(G)]_{22} - [\hbar(H)]_{22})U_a(\Phi_2(G))f_2 \end{pmatrix}$$
$$+ \begin{pmatrix} [\hbar(H)]_{11}[U_a(\Phi_2(G)) - U_a(\Phi_2(H))]f_1 + [\hbar(H)]_{12}[U_a(\Phi_2(G)) - U_a(\Phi_2(H))]f_2 \\ [\hbar(H)]_{21}[U_a(\Phi_2(G)) - U_a(\Phi_2(H))]f_1 + [\hbar(H)]_{22}[U_a(\Phi_2(G)) - U_a(\Phi_2(H))]f_2 \end{pmatrix}$$

and hence that

$$\frac{1}{4}\|T_a^{\hbar}(G)f - T_a^{\hbar}(H)f\|^2$$
$$\leqslant \frac{1}{2}\left\| \begin{pmatrix} ([\hbar(G)]_{11} - [\hbar(H)]_{11})U_a(\Phi_2(G))f_1 + ([\hbar(G)]_{12} - [\hbar(H)]_{12})U_a(\Phi_2(G))f_2 \\ ([\hbar(G)]_{21} - [\hbar(H)]_{21})U_a(\Phi_2(G))f_1 + ([\hbar(G)]_{22} - [\hbar(H)]_{22})U_a(\Phi_2(G))f_2 \end{pmatrix} \right\|^2$$
$$+ \frac{1}{2}\left\| \begin{pmatrix} [\hbar(H)]_{11}[U_a(\Phi_2(G)) - U_a(\Phi_2(H))]f_1 + [\hbar(H)]_{12}[U_a(\Phi_2(G)) - U_a(\Phi_2(H))]f_2 \\ [\hbar(H)]_{21}[U_a(\Phi_2(G)) - U_a(\Phi_2(H))]f_1 + [\hbar(H)]_{22}[U_a(\Phi_2(G)) - U_a(\Phi_2(H))]f_2 \end{pmatrix} \right\|^2$$
$$= \frac{1}{2}\|([\hbar(G)]_{11} - [\hbar(H)]_{11})U_a(\Phi_2(G))f_1 + ([\hbar(G)]_{12} - [\hbar(H)]_{12})U_a(\Phi_2(G))f_2\|_2^2$$
$$+ \frac{1}{2}\|([\hbar(G)]_{21} - [\hbar(H)]_{21})U_a(\Phi_2(G))f_1 + ([\hbar(G)]_{22} - [\hbar(H)]_{22})U_a(\Phi_2(G))f_2\|_2^2$$
$$+ \frac{1}{2}\|[\hbar(H)]_{11}[U_a(\Phi_2(G)) - U_a(\Phi_2(H))]f_1 + [\hbar(H)]_{12}[U_a(\Phi_2(G)) - U_a(\Phi_2(H))]f_2\|_2^2$$
$$+ \frac{1}{2}\|[\hbar(H)]_{21}[U_a(\Phi_2(G)) - U_a(\Phi_2(H))]f_1 + [\hbar(H)]_{22}[U_a(\Phi_2(G)) - U_a(\Phi_2(H))]f_2\|_2^2$$
$$\leqslant |[\hbar(G)]_{11} - [\hbar(H)]_{11}|^2 \cdot \|f_1\|^2 + |[\hbar(G)]_{12} - [\hbar(H)]_{12}|^2 \cdot \|f_2\|_2^2$$
$$+ |[\hbar(G)]_{21} - [\hbar(H)]_{21}|^2 \cdot \|f_1\|_2^2 + |[\hbar(G)]_{22} - [\hbar(H)]_{22}|^2 \cdot \|f_2\|_2^2$$
$$+ |[\hbar(H)]_{11}|^2 \cdot \|[U_a(\Phi_2(G)) - U_a(\Phi_2(H))]f_1\|_2^2 + |[\hbar(H)]_{12}|^2 \cdot \|[U_a(\Phi_2(G)) - U_a(\Phi_2(H))]f_2\|_2^2$$
$$+ |[\hbar(H)]_{21}|^2 \cdot \|[U_a(\Phi_2(G)) - U_a(\Phi_2(H))]f_1\|^2 + |[\hbar(H)]_{22}|^2 \cdot \|[U_a(\Phi_2(G)) - U_a(\Phi_2(H))]f_2\|_2^2 .$$

Since $\hbar$ and Φ_2 are component-wise continuous and U_a is strongly continuous, it follows that $T_a^{\hbar}$ is strongly continuous, i.e., if $G_1, G_2, \ldots$ is a sequence in $\mathrm{SL}(2, \mathbb{C})$ that converges component-wise to $G \in \mathrm{SL}(2, \mathbb{C})$, then

$$\lim_{\nu \to \infty} \|[T_a^{\hbar}(G_\nu) - T_a^{\hbar}(G)]f\| = 0 ,$$

for every $f \in (L^2_{\mathbb{C}}(\mathbb{R}^3, \varphi_a))^2$.

Hence, we arrive at the following result. If $a \geqslant 0$ and $\hbar \in \{\hbar_R, \hbar_L\}$, then

$$T_a^{\hbar} : \mathrm{SL}(2,\mathbb{C}) \to \mathrm{L}((L^2_{\mathbb{C}}(\mathbb{R}^3,\varphi_a))^2, (L^2_{\mathbb{C}}(\mathbb{R}^3,\varphi_a))^2) \, ,$$

defined by

$$\begin{aligned} T_a^{\hbar}(G)\begin{pmatrix} f_1 \\ f_2 \end{pmatrix} &:= \hbar(G)\cdot\begin{pmatrix} U_a(\Phi_2(G))f_1 \\ U_a(\Phi_2(G))f_2 \end{pmatrix} \\ &= \begin{pmatrix} [\hbar(G)]_{11}U_a(\Phi_2(G))f_1 + [\hbar(G)]_{12}U_a(\Phi_2(G))f_2 \\ [\hbar(G)]_{21}U_a(\Phi_2(G))f_1 + [\hbar(G)]_{22}U_a(\Phi_2(G))f_2 \end{pmatrix} , \end{aligned}$$

for every $G \in \mathrm{SL}(2,\mathbb{C})$ and ${}^t(f_1, f_2) \in (L^2_{\mathbb{C}}(\mathbb{R}^3,\varphi_a))^2$, is a representation of $\mathrm{SL}(2,\mathbb{C})$. In addition, $T_a^{\hbar}$ is strongly continuous, i.e., if $G_1, G_2, \dots$ is a sequence in $\mathrm{SL}(2,\mathbb{C})$ that converges component-wise to $G \in \mathrm{SL}(2,\mathbb{C})$, then

$$\lim_{\nu\to\infty} \|[T_a^{\hbar}(G_\nu) - T_a^{\hbar}(G)]f\| = 0 \, ,$$

for every $f \in (L^2_{\mathbb{C}}(\mathbb{R}^3,\varphi_a))^2$. Further, $T_a^{\hbar_R}$ and $T_a^{\hbar_L}$ coincide on $\mathrm{SU}(2)$ and $T_a^{\hbar_R}(G)$ and $T_a^{\hbar_L}(G)$ are unitary for every $G \in \mathrm{SU}(2)$.

1.4.1 Generators Associated with Rotations and Lorentz Boosts

Further, if

$$G : (\mathbb{R}, +) \to \mathrm{SL}(2,\mathbb{C})$$

is a one-parameter group, i.e., such that

$$G(s_1 + s_2) = G(s_1)\cdot G(s_2) \; , \; G(0) = E \; ,$$

for all $s_1, s_2 \in \mathbb{R}$ and such that, for every sequence $s_1, s_2, \dots$ in $\mathbb{R}$ that is convergent to $s \in \mathbb{R}$, the corresponding sequence $G(s_1), G(s_2), \dots$ converges component-wise to $G(s)$, then $T_a^{\hbar} \circ G$ is a strongly continuous one-parameter group. Hence, there is an infinitesimal generator[1] $A_G^{\hbar}$ in

$$\boxed{X := (L^2_{\mathbb{C}}(\mathbb{R}^3,\varphi_a))^2 \, ,}$$

given by

$$D(A_G^{\hbar}) = \{f \in X : \lim_{s\to 0, s\neq 0} \frac{1}{s}\left[(T_a^{\hbar}\circ G)(s) - \mathrm{id}_X\right] f \text{ exists}\}$$

[1] E.g., see [4], Sect. 4.5.

and for every $f \in D(A_G^{\hbar})$ by[2]

$$A_G^{\hbar} f = \frac{1}{i} \lim_{s \to 0, s \neq 0} \frac{1}{s} \left[(T_a^{\hbar} \circ G)(s) - \mathrm{id}_X \right] f \ .$$

1.4.1.1 Generators Associated with Rotations

In particular, since $\mathrm{Ran}(G) \subset \mathrm{SU}(2)$, $T_a^{\hbar} \circ G$ is a strongly continuous one-parameter unitary group and, since

$$\hbar_R(H) = \hbar_L(H) = H \ ,$$

for every $H \in \mathrm{SU}(2)$, we have that

$$T_a^{\hbar} \circ G = T_a^{\hbar_R} \circ G \ .$$

Then, according to Stone's theorem,

$$\exp(is A_G^{\hbar_R}) = (T_a^{\hbar_R} \circ G)(s) \ ,$$

for every $s \in \mathbb{R}$, where $A_G^{\hbar_R}$ is (densely-defined, linear and) self-adjoint.

Further, according to (1.5) and (1.8), we have that

$$\Phi_2\left(\begin{pmatrix} \cos(\theta/2) & i\sin(\theta/2) \\ i\sin(\theta/2) & \cos(\theta/2) \end{pmatrix}\right) = \begin{pmatrix} 1 & 0 & 0 & 0 \\ 0 & 1 & 0 & 0 \\ 0 & 0 & \cos(\theta) & \sin(\theta) \\ 0 & 0 & -\sin(\theta) & \cos(\theta) \end{pmatrix} = M_1(\theta) \ ,$$

$$\Phi_2\left(\begin{pmatrix} \cos(\varphi/2) & \sin(\varphi/2) \\ -\sin(\varphi/2) & \cos(\varphi/2) \end{pmatrix}\right) = \begin{pmatrix} 1 & 0 & 0 & 0 \\ 0 & \cos(\varphi) & 0 & -\sin(\varphi) \\ 0 & 0 & 1 & 0 \\ 0 & \sin(\varphi) & 0 & \cos(\varphi) \end{pmatrix} = M_2(\varphi) \ ,$$

$$\Phi_2\left(\begin{pmatrix} e^{i\psi/2} & 0 \\ 0 & e^{-i\psi/2} \end{pmatrix}\right) = \begin{pmatrix} 1 & 0 & 0 & 0 \\ 0 & \cos(\psi) & \sin(\psi) & 0 \\ 0 & -\sin(\psi) & \cos(\psi) & 0 \\ 0 & 0 & 0 & 1 \end{pmatrix} = M_3(\psi) \ ,$$

for $\theta, \varphi, \psi \in \mathbb{R}$, where the one-parameter groups of rotations about the coordinate axes, $M_j : \mathbb{R} \to \mathcal{L}_+^{\uparrow}$, $j \in \{1, 2, 3\}$, are defined by

[2] The following definition is not standard. In a standard definition, the factor $1/i$ in the following definition is replaced by -1 or 1, see [4], Sect. 4.5. Hence, in the sense of strongly semigroups of operators, the generator of $T_a^{\hbar} \circ G$ is given by $-iA_G^{\hbar}$.

$$M_1(s) := \begin{pmatrix} 1 & 0 & 0 & 0 \\ 0 & 1 & 0 & 0 \\ 0 & 0 & \cos(s) & \sin(s) \\ 0 & 0 & -\sin(s) & \cos(s) \end{pmatrix} ,$$

$$M_2(s) := \begin{pmatrix} 1 & 0 & 0 & 0 \\ 0 & \cos(s) & 0 & -\sin(s) \\ 0 & 0 & 1 & 0 \\ 0 & \sin(s) & 0 & \cos(s) \end{pmatrix} ,$$

$$M_3(s) := \begin{pmatrix} 1 & 0 & 0 & 0 \\ 0 & \cos(s) & \sin(s) & 0 \\ 0 & -\sin(s) & \cos(s) & 0 \\ 0 & 0 & 0 & 1 \end{pmatrix} , \tag{1.17}$$

for every $s \in \mathbb{R}$.

Exercise 1.4 For every $s \in \mathbb{C}$ and $j \in \{1, 2, 3\}$, show that

$$\exp(s\sigma_j) = \cosh(s).E + \sinh(s).\sigma_j$$

and hence that

$$\exp(is\sigma_1) = \begin{pmatrix} \cos(s) & i\sin(s) \\ i\sin(s) & \cos(s) \end{pmatrix} , \quad \exp(is\sigma_2) = \begin{pmatrix} \cos(s) & \sin(s) \\ -\sin(s) & \cos(s) \end{pmatrix} ,$$

$$\exp(is\sigma_3) = \begin{pmatrix} e^{is} & 0 \\ 0 & e^{-is} \end{pmatrix} ,$$

$$\exp(s\sigma_1) = \begin{pmatrix} \cosh(s) & \sinh(s) \\ \sinh(s) & \cosh(s) \end{pmatrix} , \quad \exp(s\sigma_2) = \begin{pmatrix} \cosh(s) & -i\sinh(s) \\ i\sinh(s) & \cosh(s) \end{pmatrix} ,$$

$$\exp(s\sigma_3) = \begin{pmatrix} e^{s} & 0 \\ 0 & e^{-s} \end{pmatrix} ,$$

for every $s \in \mathbb{C}$.

From Exercise 1.4, it follows that

$$\Phi_2(G_j(\varphi)) = M_j(\varphi) ,$$

for $j \in \{1, 2, 3\}$ and $\varphi \in \mathbb{R}$, where the $M_1, M_2, M_3 : \mathbb{R} \to \mathcal{L}_+^\uparrow$ are defined by (1.17) and the one-parameter groups $G_1, G_2, G_3 : \mathbb{R} \to \mathrm{SU}(2)$ are defined by

$$G_1(\varphi) := \exp(i\,(\varphi/2)\,\sigma_1) , \quad G_2(\varphi) := \exp(i\,(\varphi/2)\,\sigma_2) ,$$
$$G_3(\varphi) := \exp(i\,(\varphi/2)\,\sigma_3) ,$$

for every $\varphi \in \mathbb{R}$. From the continuity of the exponential function, it follows that G_j is continuous, for every $j \in \{1, 2, 3\}$. Further, since $\hbar(H) = H$, for every $H \in \mathrm{SU}(2)$, we have that

$$\hbar(G_j(\varphi)) = G_j(\varphi) = \exp(i\,(\varphi/2)\,\sigma_j)\ ,$$

for $j \in \{1, 2, 3\}$ and every $\varphi \in \mathbb{R}$. Hence,

$$\begin{aligned}
(T_a^{\hbar_R} \circ G_j)(\varphi) f &= \hbar_R(G_j(\varphi)) \cdot \begin{pmatrix} U_a(\Phi_2(G_j(\varphi))) f_1 \\ U_a(\Phi_2(G_j(\varphi))) f_2 \end{pmatrix} \\
&= G_j(\varphi) \cdot \begin{pmatrix} U_a(M_j(\varphi)) f_1 \\ U_a(M_j(\varphi)) f_2 \end{pmatrix} = G_j(\varphi) \cdot \begin{pmatrix} \exp\left(\frac{i\varphi}{\hbar}\,\hat{L}_j\right) f_1 \\ \exp\left(\frac{i\varphi}{\hbar}\,\hat{L}_j\right) f_2 \end{pmatrix} ,
\end{aligned}$$

for every $\varphi \in \mathbb{R}$ and $f \in X$, where Formula 5.19 of Part I has been used. Further, for $f \in X$ and $\varphi \in \mathbb{R}^*$, it follows that

$$\begin{aligned}
&\frac{1}{i\varphi}\left[(T_a^{\hbar_R} \circ G_j)(\varphi) - \mathrm{id}_X\right] f = \frac{1}{i\varphi}\left[G_j(\varphi) \cdot \begin{pmatrix} \exp(\frac{i\varphi}{\hbar}\,\hat{L}_j) f_1 \\ \exp(\frac{i\varphi}{\hbar}\,\hat{L}_j) f_2 \end{pmatrix} - \begin{pmatrix} f_1 \\ f_2 \end{pmatrix}\right] \\
&= G_j(\varphi) \cdot \begin{pmatrix} (i\varphi)^{-1}\,[\exp(\frac{i\varphi}{\hbar}\,\hat{L}_j) - 1] f_1 \\ (i\varphi)^{-1}\,[\exp(\frac{i\varphi}{\hbar}\,\hat{L}_j) - 1] f_2 \end{pmatrix} + \frac{1}{i\varphi}\,[G_j(\varphi) - 1] \begin{pmatrix} f_1 \\ f_2 \end{pmatrix} \\
&= \begin{pmatrix} (i\varphi)^{-1}\,[\exp(\frac{i\varphi}{\hbar}\,\hat{L}_j) - 1] f_1 \\ (i\varphi)^{-1}\,[\exp(\frac{i\varphi}{\hbar}\,\hat{L}_j) - 1] f_2 \end{pmatrix} + \frac{1}{2}\sigma_j \begin{pmatrix} f_1 \\ f_2 \end{pmatrix} \\
&\quad + [G_j(\varphi) - 1] \cdot \begin{pmatrix} (i\varphi)^{-1}\,[\exp(\frac{i\varphi}{\hbar}\,\hat{L}_j) - 1] f_1 \\ (i\varphi)^{-1}\,[\exp(\frac{i\varphi}{\hbar}\,\hat{L}_j) - 1] f_2 \end{pmatrix} \\
&\quad + \left\{\frac{1}{i\varphi}\,[G_j(\varphi) - 1] - \frac{1}{2}\sigma_j\right\} \begin{pmatrix} f_1 \\ f_2 \end{pmatrix} .
\end{aligned}$$

We note that

$$\begin{aligned}
&\|G_j(\varphi) - 1\|_{\mathrm{op}} = \|\exp(\frac{i\varphi}{2}\,\sigma_j) - 1\|_{\mathrm{op}} = \|\sum_{k=1}^{\infty} \frac{1}{k!}\,(\frac{i\varphi}{2}\,\sigma_j)^k\,\|_{\mathrm{op}} \\
&\leqslant \sum_{k=1}^{\infty} \frac{1}{k!}\left(\frac{|\varphi|}{2}\right)^k = \frac{|\varphi|}{2} \sum_{k=1}^{\infty} \frac{1}{k!}\left(\frac{|\varphi|}{2}\right)^{k-1} \leqslant \frac{|\varphi|}{2} \sum_{k=1}^{\infty} \frac{1}{(k-1)!}\left(\frac{|\varphi|}{2}\right)^{k-1} \\
&= \frac{|\varphi|}{2}\,\exp\left(\frac{|\varphi|}{2}\right) , \\
&\|\frac{1}{\varphi}\,[G_j(\varphi) - 1] - \frac{i}{2}\,\sigma_j\|_{\mathrm{op}} = \|\frac{1}{\varphi}\,[\exp(\frac{i\varphi}{2}\,\sigma_j) - 1] - \frac{i}{2}\,\sigma_j\|_{\mathrm{op}} \\
&= \|\frac{1}{\varphi}\left[\sum_{k=1}^{\infty} \frac{1}{k!}\,(\frac{i\varphi}{2}\,\sigma_j)^k\right] - \frac{i}{2}\,\sigma_j\|_{\mathrm{op}} = \|\frac{1}{\varphi}\left[\sum_{k=2}^{\infty} \frac{1}{k!}\,(\frac{i\varphi}{2}\,\sigma_j)^k\right]\|_{\mathrm{op}}
\end{aligned}$$

$$\leqslant \frac{1}{|\varphi|} \sum_{k=2}^{\infty} \frac{1}{k!} \left(\frac{|\varphi|}{2}\right)^k = \frac{|\varphi|}{4} \sum_{k=2}^{\infty} \frac{1}{k!} \left(\frac{|\varphi|}{2}\right)^{k-2} \leqslant \frac{|\varphi|}{4} \sum_{k=2}^{\infty} \frac{1}{(k-2)!} \left(\frac{|\varphi|}{2}\right)^{k-2}$$
$$= \frac{|\varphi|}{4} \exp\left(\frac{|\varphi|}{2}\right) ,$$

where we used that $\|\sigma_j\|_{\text{op}} = 1$, since

$$|\sigma_1 \cdot \begin{pmatrix} z_1 \\ z_2 \end{pmatrix}| = |\begin{pmatrix} 0 & 1 \\ 1 & 0 \end{pmatrix} \cdot \begin{pmatrix} z_1 \\ z_2 \end{pmatrix}| = |\begin{pmatrix} z_2 \\ z_1 \end{pmatrix}| = |\begin{pmatrix} z_1 \\ z_2 \end{pmatrix}| ,$$
$$|\sigma_2 \cdot \begin{pmatrix} z_1 \\ z_2 \end{pmatrix}| = |\begin{pmatrix} 0 & -i \\ i & 0 \end{pmatrix} \cdot \begin{pmatrix} z_1 \\ z_2 \end{pmatrix}| = |\begin{pmatrix} -iz_2 \\ iz_1 \end{pmatrix}| = |\begin{pmatrix} z_1 \\ z_2 \end{pmatrix}| ,$$
$$|\sigma_3 \cdot \begin{pmatrix} z_1 \\ z_2 \end{pmatrix}| = |\begin{pmatrix} 1 & 0 \\ 0 & -1 \end{pmatrix} \cdot \begin{pmatrix} z_1 \\ z_2 \end{pmatrix}| = |\begin{pmatrix} z_1 \\ -z_2 \end{pmatrix}| = |\begin{pmatrix} z_1 \\ z_2 \end{pmatrix}| ,$$

for every ${}^t(z_1, z_2) \in \mathbb{C}^2$, where $|\ |$ denotes the canonical norm for $\mathbb{C}^2$. We note that the coordinate projections $p_1, p_2 : X \to L^2_{\mathbb{C}}(\mathbb{R}^3, \varphi_a)$ as well as the inclusions $\iota_1, \iota_2 : L^2_{\mathbb{C}}(\mathbb{R}^3, \varphi_a) \hookrightarrow X$, defined by $p_1 f := f_1$, $p_2 f := f$, for every $f = {}^t(f_1, f_2) \in X$ and $\iota_1 f := {}^t(f, 0)$ and $\iota_2 f := {}^t(0, f)$, for every $f \in L^2_{\mathbb{C}}(\mathbb{R}^3, \varphi_a)$ are linear and continuous, since

$$\|p_k f\|_2 = \|f_k\|_2 \leqslant \|f\| \ , \ \|\iota_k g\| = \|g\|_2 \ ,$$

for every $f = {}^t(f_1, f_2) \in X, g \in L^2_{\mathbb{C}}(\mathbb{R}^3, \varphi_a)$ and $k \in \{1, 2\}$. As a consequence, an element $f \in X$ is part of the domain $D(A^{\hbar_R}_{G_j})$ of $A^{\hbar_R}_{G_j}$ if and only if $f \in D(\hat{L}_j) \times D(\hat{L}_j)$ and if $f \in D(A^{\hbar_R}_{G_j})$, then

$$A^{\hbar_R}_{G_j} f = \frac{1}{\hbar} \begin{pmatrix} \hat{L}_j f_1 \\ \hat{L}_j f_2 \end{pmatrix} + \frac{1}{2} \sigma_j \cdot \begin{pmatrix} f_1 \\ f_2 \end{pmatrix} .$$

Hence, for $j \in \{1, 2, 3\}$, we define the, densely-defined, linear and self-adjoint, j-th component $\hat{J}_j : D(\hat{L}_j) \times D(\hat{L}_j) \to X$ of total angular momentum by

$$\hat{J}_j f := \hbar A^{\hbar_R}_{G_j} f = \begin{pmatrix} \hat{L}_j f_1 \\ \hat{L}_j f_2 \end{pmatrix} + \frac{\hbar}{2} \sigma_j \cdot \begin{pmatrix} f_1 \\ f_2 \end{pmatrix} , \tag{1.18}$$

for every $f = {}^t(f_1, f_2) \in D(\hat{L}_j) \times D(\hat{L}_j)$ as well as the, bounded linear and self-adjoint, j-th component $\hat{S}_j : X \to X$ of intrinsic angular momentum by

$$\hat{S}_j f := \frac{\hbar}{2} \sigma_j \cdot \begin{pmatrix} f_1 \\ f_2 \end{pmatrix} ,$$

for every $f = {}^t(f_1, f_2) \in X$, where $X = (L^2_{\mathbb{C}}(\mathbb{R}^3, \varphi_a))^2$.

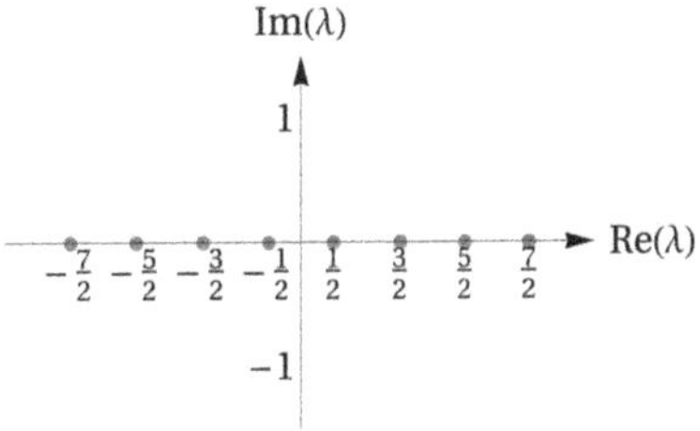

Fig. 1.4 Depiction of the spectrum (in red) of $\hat{J}_3$

Exercise 1.5 Show that the operator $\hat{J}_3$ has a pure point spectrum, consisting of eigenvalues of infinite multiplicity, given by (Fig. 1.4)

$$\hbar\left(\mathbb{Z}+\frac{1}{2}\right) .$$

Further, since according to Stone's theorem, we have that

$$\begin{aligned}\exp\Big(i\,\frac{\varphi}{\hbar}\,\hat{J}_j\Big)\,f &= \exp(i\varphi A^{\hbar_R}_{G_j})f = (T^{\hbar_R}_a \circ G_j)(\varphi)f \\ &= G_j(\varphi)\cdot\begin{pmatrix}\exp\Big(\frac{i\varphi}{\hbar}\,\hat{L}_j\Big)\,f_1\\ \exp\Big(\frac{i\varphi}{\hbar}\,\hat{L}_j\Big)\,f_2\end{pmatrix} ,\end{aligned}$$

it follows that

$$\exp\Big(\frac{i\varphi}{\hbar}\,\hat{J}_j\Big)f = G_j(\varphi)\cdot\begin{pmatrix}\exp(\frac{i\varphi}{\hbar}\,\hat{L}_j)f_1\\ \exp(\frac{i\varphi}{\hbar}\,\hat{L}_j)f_2\end{pmatrix} , \tag{1.19}$$

for every $f = {}^t(f_1, f_2) \in X$ and $\varphi \in \mathbb{R}$.

Also, we note that the eigenvalues of $\hat{S}_j$ are given by $-\hbar/2$ and $\hbar/2$ as well as that

$$\begin{aligned}
&\ker\big(\hat{S}_1+\frac{\hbar}{2}\big) = L^2_{\mathbb{C}}(\mathbb{R}^3,\varphi_a)\,.\begin{pmatrix}1\\-1\end{pmatrix} , \quad \ker\big(\hat{S}_1-\frac{\hbar}{2}\big) = L^2_{\mathbb{C}}(\mathbb{R}^3,\varphi_a)\,.\begin{pmatrix}1\\1\end{pmatrix} ,\\
&\ker\big(\hat{S}_2+\frac{\hbar}{2}\big) = L^2_{\mathbb{C}}(\mathbb{R}^3,\varphi_a)\,.\begin{pmatrix}1\\-i\end{pmatrix} , \quad \ker\big(\hat{S}_2-\frac{\hbar}{2}\big) = L^2_{\mathbb{C}}(\mathbb{R}^3,\varphi_a)\,.\begin{pmatrix}1\\i\end{pmatrix} ,\\
&\ker\big(\hat{S}_3+\frac{\hbar}{2}\big) = L^2_{\mathbb{C}}(\mathbb{R}^3,\varphi_a)\,.\begin{pmatrix}0\\1\end{pmatrix} , \quad \ker\big(\hat{S}_3-\frac{\hbar}{2}\big) = L^2_{\mathbb{C}}(\mathbb{R}^3,\varphi_a)\,.\begin{pmatrix}1\\0\end{pmatrix} .
\end{aligned}$$

Hence, we obtain the following result.

For $j \in \{1, 2, 3\}$, there is a Hilbert basis of X, consisting of eigenvectors of $\hat{S}_j$, corresponding to the eigenvalues $-\hbar/2$ and $\hbar/2$. Therefore, $\hat{S}_j$ has a pure point spectrum given by

$$\sigma(\hat{S}_j) = \left\{-\frac{\hbar}{2}, \frac{\hbar}{2}\right\} .$$

In addition, we have that

$$\begin{aligned}
\exp\Big(\frac{i\varphi}{\hbar}\,\hat{S}_j\Big)f &= \sum_{k=0}^{\infty}\frac{1}{k!}\Big(\frac{i\varphi}{\hbar}\,\hat{S}_j\Big)^k f = \sum_{k=0}^{\infty}\frac{1}{k!}\Big(\frac{i\varphi}{2}\,\sigma_j\Big)^k f \\
&= \sum_{k=0}^{\infty}\frac{1}{(2k)!}\Big(\frac{i\varphi}{2}\Big)^{2k}\sigma_j^{2k} f + \sum_{k=0}^{\infty}\frac{1}{(2k+1)!}\Big(\frac{i\varphi}{2}\Big)^{2k+1}\sigma_j^{2k+1} f \\
&= \sum_{k=0}^{\infty}\frac{1}{(2k)!}\Big(\frac{i\varphi}{2}\Big)^{2k} f + \sum_{k=0}^{\infty}\frac{1}{(2k+1)!}\Big(\frac{i\varphi}{2}\Big)^{2k+1}\sigma_j f \\
&= \cosh\Big(\frac{i\varphi}{2}\Big)f + \sinh\Big(\frac{i\varphi}{2}\Big)\sigma_j f = \cos\Big(\frac{\varphi}{2}\Big)f + i\sin\Big(\frac{\varphi}{2}\Big)\,\sigma_j f ,
\end{aligned}$$

for every $f \in X$ and $\varphi \in \mathbb{R}$, where we used that

$$\sigma_j^{2k} = E \ , \ \sigma_j^{2k+1} = \sigma_j \ ,$$

for every $k \in \mathbb{N}$. Hence,

$$\boxed{\exp\Big(\frac{i\varphi}{\hbar}\,\hat{S}_j\Big)f = \cos\Big(\frac{\varphi}{2}\Big)f + i\sin\Big(\frac{\varphi}{2}\Big)\,\sigma_j f ,}$$

for every $f \in X$ and $\varphi \in \mathbb{R}$. As a consequence,

$$\exp\Big(\frac{i\,(\varphi + 2k\pi)}{\hbar}\,\hat{S}_j\Big)f = (-1)^k \exp\Big(\frac{i\varphi}{\hbar}\,\hat{S}_j\Big)f ,$$

for every $f \in X$, $\varphi \in \mathbb{R}$ and $k \in \mathbb{Z}$, i.e., an increase of the angle φ about $2k\pi$, $k \in \mathbb{Z}$, results in a multiplication by a phase factor.

1.4.1.2 Generators Associated with Lorentz Boosts

In the following, we consider the generators that are associated with one-parameter subgroups of the restricted Lorentz group that are associated with Lorentz boosts. We note that it follows from Exercise 1.4, (1.5) and (1.9) that

$$\Phi_2(G_j(s)) = M_{0j}(s) ,$$

for $j \in \{1, 2, 3\}$ and $s \in \mathbb{R}$, where the one-parameter groups of Lorentz boosts, M_{01}, M_{02}, $M_{03} : \mathbb{R} \to \mathcal{L}_+^\uparrow$, are defined by

$$M_{01}(s) := \begin{pmatrix} \cosh(s) & \sinh(s) & 0 & 0 \\ \sinh(s) & \cosh(s) & 0 & 0 \\ 0 & 0 & 1 & 0 \\ 0 & 0 & 0 & 1 \end{pmatrix} ,$$

$$M_{02}(s) := \begin{pmatrix} \cosh(s) & 0 & \sinh(s) & 0 \\ 0 & 1 & 0 & 0 \\ \sinh(s) & 0 & \cosh(s) & 0 \\ 0 & 0 & 0 & 1 \end{pmatrix} ,$$

$$M_{03}(s) := \begin{pmatrix} \cosh(s) & 0 & 0 & \sinh(s) \\ 0 & 1 & 0 & 0 \\ 0 & 0 & 1 & 0 \\ \sinh(s) & 0 & 0 & \cosh(s) \end{pmatrix} , \tag{1.20}$$

for every $s \in \mathbb{R}$, and the one-parameter groups $G_1, G_2, G_3 : \mathbb{R} \to \mathrm{SL}(2, \mathbb{C})$ are defined by

$$G_1(s) := \exp((s/2)\,\sigma_1) , \quad G_2(s) := \exp((s/2)\,\sigma_2) ,$$
$$G_3(s) := \exp((s/2)\,\sigma_3) ,$$

for every $s \in \mathbb{R}$. From the continuity of the exponential function, it follows that G_1, G_2 and G_3 are continuous. We note that

$$\hbar_R(G_j(s)) = G_j(s) = \exp((s/2)\,\sigma_j) ,$$
$$\hbar_L(G_j(s)) = [(G_j(s))^*]^{-1} = \exp(-(s/2)\,\sigma_j) ,$$

for $j \in \{1, 2, 3\}$ and every $s \in \mathbb{R}$. Hence,

$$(T_a^{\hbar} \circ G_j)(s) = \hbar(G_j(s)) \cdot \begin{pmatrix} U_a(\Phi_2(G_j(s)))\, f_1 \\ U_a(\Phi_2(G_j(s))\, f_2 \end{pmatrix}$$

$$= \hbar(G_j(s)) \cdot \begin{pmatrix} U_a(M_{0j}(s))\, f_1 \\ U_a(M_{0j}(s))\, f_2 \end{pmatrix} = \hbar(G_j(s)) \cdot \begin{pmatrix} \exp\left(i\,\frac{s}{\hbar}\,\hat{L}_{0j}\right) f_1 \\ \exp\left(i\,\frac{s}{\hbar}\,\hat{L}_{0j}\right) f_2 \end{pmatrix} ,$$

for every $s \in \mathbb{R}$ and $f \in X$, where Formula 5.21 of Part I has been used. Further, for $f \in X$ and $s \in \mathbb{R}^*$, it follows that

$$\frac{1}{s}\left[(T_a^{\hbar} \circ G_j)(s) - \mathrm{id}_X\right] f = \frac{1}{s}\left[\hbar(G_j(s)) \cdot \begin{pmatrix} \exp(i\,\frac{s}{\hbar}\,\hat{L}_j) f_1 \\ \exp(i\,\frac{s}{\hbar}\,\hat{L}_j) f_2 \end{pmatrix} - \begin{pmatrix} f_1 \\ f_2 \end{pmatrix}\right]$$

$$= \hbar(G_j(s)) \cdot \begin{pmatrix} s^{-1}\,[\exp(i\,\frac{s}{\hbar}\,\hat{L}_j) - 1] f_1 \\ s^{-1}\,[\exp(i\,\frac{s}{\hbar}\,\hat{L}_j) - 1] f_2 \end{pmatrix} + \frac{1}{s}\,[\hbar(G_j(s)) - 1] \begin{pmatrix} f_1 \\ f_2 \end{pmatrix}$$

$$= \begin{pmatrix} s^{-1}\,[\exp(i\,\frac{s}{\hbar}\,\hat{L}_j) - 1] f_1 \\ s^{-1}\,[\exp(i\,\frac{s}{\hbar}\,\hat{L}_j) - 1] f_2 \end{pmatrix} \pm \frac{1}{2}\sigma_j \begin{pmatrix} f_1 \\ f_2 \end{pmatrix}$$

$$+ [\hbar(G_j(s)) - 1] \cdot \begin{pmatrix} s^{-1}\,[\exp(i\,\frac{s}{\hbar}\,\hat{L}_j) - 1] f_1 \\ s^{-1}\,[\exp(i\,\frac{s}{\hbar}\,\hat{L}_j) - 1] f_2 \end{pmatrix}$$

$$+ \left\{\frac{1}{s}\,[\hbar(G_j(s)) - 1] \mp \frac{1}{2}\sigma_j\right\} \begin{pmatrix} f_1 \\ f_2 \end{pmatrix} ,$$

where here and in the following the upper signs in $\pm$ and $\mp$ refer to $\hbar = \hbar_R$ and the lower signs in $\pm$ and $\mp$ refer to $\hbar = \hbar_L$. We note that

$$\|\hbar(G_j(s)) - 1\|_{\mathrm{op}} = \|\exp(\pm\frac{s}{2}\sigma_j) - 1\|_{\mathrm{op}} = \|\sum_{k=1}^{\infty} \frac{1}{k!}\,(\pm\frac{s}{2}\sigma_j)^k\|_{\mathrm{op}}$$

$$\leqslant \sum_{k=1}^{\infty} \frac{1}{k!}\left(\frac{|s|}{2}\right)^k = \frac{|s|}{2}\sum_{k=1}^{\infty} \frac{1}{k!}\left(\frac{|s|}{2}\right)^{k-1} \leqslant \frac{|s|}{2}\sum_{k=1}^{\infty} \frac{1}{(k-1)!}\left(\frac{|s|}{2}\right)^{k-1}$$

$$= \frac{|s|}{2}\exp\left(\frac{|s|}{2}\right) ,$$

$$\|\frac{1}{s}\,[\hbar(G_j(s)) - 1] \mp \frac{1}{2}\sigma_j\|_{\mathrm{op}} = \|\frac{1}{s}\,[\exp(\pm\frac{s}{2}\sigma_j) - 1] \mp \frac{1}{2}\sigma_j\|_{\mathrm{op}}$$

$$= \|\frac{1}{s}\left[\sum_{k=1}^{\infty} \frac{1}{k!}\,(\pm\frac{s}{2}\sigma_j)^k\right] \mp \frac{1}{2}\sigma_j\|_{\mathrm{op}} = \|\frac{1}{s}\left[\sum_{k=2}^{\infty} \frac{1}{k!}\,(\pm\frac{s}{2}\sigma_j)^k\right]\|_{\mathrm{op}}$$

$$\leqslant \frac{1}{|s|}\sum_{k=2}^{\infty} \frac{1}{k!}\left(\frac{|s|}{2}\right)^k = \frac{|s|}{4}\sum_{k=2}^{\infty} \frac{1}{k!}\left(\frac{|s|}{2}\right)^{k-2} \leqslant \frac{|s|}{4}\sum_{k=2}^{\infty} \frac{1}{(k-2)!}\left(\frac{|s|}{2}\right)^{k-2}$$

$$= \frac{|s|}{4}\exp\left(\frac{|s|}{2}\right) ,$$

where we used that $\|\sigma_j\|_{\mathrm{op}} = 1$, since

$$|\sigma_1 \cdot \begin{pmatrix} z_1 \\ z_2 \end{pmatrix}| = |\begin{pmatrix} 0 & 1 \\ 1 & 0 \end{pmatrix} \cdot \begin{pmatrix} z_1 \\ z_2 \end{pmatrix}| = |\begin{pmatrix} z_2 \\ z_1 \end{pmatrix}| = |\begin{pmatrix} z_1 \\ z_2 \end{pmatrix}| ,$$

$$|\sigma_2 \cdot \begin{pmatrix} z_1 \\ z_2 \end{pmatrix}| = |\begin{pmatrix} 0 & -i \\ i & 0 \end{pmatrix} \cdot \begin{pmatrix} z_1 \\ z_2 \end{pmatrix}| = |\begin{pmatrix} -iz_2 \\ iz_1 \end{pmatrix}| = |\begin{pmatrix} z_1 \\ z_2 \end{pmatrix}| ,$$

$$|\sigma_3 \cdot \begin{pmatrix} z_1 \\ z_2 \end{pmatrix}| = |\begin{pmatrix} 1 & 0 \\ 0 & -1 \end{pmatrix} \cdot \begin{pmatrix} z_1 \\ z_2 \end{pmatrix}| = |\begin{pmatrix} z_1 \\ -z_2 \end{pmatrix}| = |\begin{pmatrix} z_1 \\ z_2 \end{pmatrix}| ,$$

for every ${}^t(z_1, z_2) \in \mathbb{C}^2$, where $|\ |$ denotes the canonical norm for $\mathbb{C}^2$. We note that the coordinate projections $p_1, p_2 : X \to L^2_{\mathbb{C}}(\mathbb{R}^3, \varphi_a)$ as well as the inclusions $\iota_1, \iota_2 : L^2_{\mathbb{C}}(\mathbb{R}^3, \varphi_a) \to X$, defined by $p_1 f := f_1$, $p_2 f := f$, for every $f \in X$ and $\iota_1 f := {}^t(f, 0)$ and $\iota_2 f := {}^t(0, f)$, for every $f \in L^2_{\mathbb{C}}(\mathbb{R}^3, \varphi_a)$ are linear and continuous, since

$$\|p_k f\|_2 = \|f_k\|_2 \leqslant \|f\| \ , \ \ \|\iota_k g\| = \|g\|_2 \ ,$$

for every $f \in X$, $g \in L^2_{\mathbb{C}}(\mathbb{R}^3, \varphi_a)$ and $k \in \{1, 2\}$. As a consequence, an element $f \in X$ is part of the domain $D(A^{\hbar}_{G_j})$ of $A^{\hbar}_{G_j}$ if and only if $f \in D(\hat{L}_{0j}) \times D(\hat{L}_{0j})$ and if $f \in D(A^{\hbar}_{G_j})$, then

$$A^{\hbar}_{G_j} f = -\frac{i}{\hbar} \begin{pmatrix} \hat{L}_{0j} f_1 \\ \hat{L}_{0j} f_2 \end{pmatrix} \mp \frac{1}{2} \sigma_j \cdot \begin{pmatrix} f_1 \\ f_2 \end{pmatrix} .$$

Hence, for $j \in \{1, 2, 3\}$, by

$$\hat{J}_{R0j} f := i\hbar A^{\hbar_R}_{G_j} f = \begin{pmatrix} \hat{L}_{0j} f_1 \\ \hat{L}_{0j} f_2 \end{pmatrix} - \frac{\hbar}{2} i\sigma_j \cdot \begin{pmatrix} f_1 \\ f_2 \end{pmatrix} ,$$

$$\hat{J}_{L0j} f := i\hbar A^{\hbar_L}_{G_j} f = \begin{pmatrix} \hat{L}_{0j} f_1 \\ \hat{L}_{0j} f_2 \end{pmatrix} + \frac{\hbar}{2} i\sigma_j \cdot \begin{pmatrix} f_1 \\ f_2 \end{pmatrix} , \tag{1.21}$$

for every $f \in D(\hat{L}_{0j}) \times D(\hat{L}_{0j})$, there are given densely-defined, linear operators in $(L^2_{\mathbb{C}}(\mathbb{R}^3, \varphi_a))^2$, such that $-(i/\hbar)\hat{J}_{R0j}$ and $-(i/\hbar)\hat{J}_{L0j}$ are generators of strongly continuous one-parameter groups on $(L^2_{\mathbb{C}}(\mathbb{R}^3, \varphi_a))^2$.

Exercise 1.6 Show that the spectra of $\hat{J}_{R03}$ and $\hat{J}_{L03}$ are given by

$$\hbar\left(\frac{i}{2} + \mathbb{R}\right) \cup \hbar\left(-\frac{i}{2} + \mathbb{R}\right) .$$

1.5 A Semi-direct Product of $\mathbb{R}^4$ and SL(2, $\mathbb{C}$) and a Double Cover of $\mathcal{P}_+^\uparrow$

In the following, let $a \geqslant 0$ and $\hbar \in \{\hbar_R, \hbar_L\}$. Our subsequent goal is the extension of $T^{\hbar}_a$ to a representation $\hat{T}^{\hbar}_a$ that includes the translation group. For this purpose, in the next step, we are going to define a semi-direct product of the translation group and SL(2, $\mathbb{C}$). With the help of the covering Φ_2, we define a multiplication $\cdot : (\mathbb{R}^4 \times \text{SL}(2, \mathbb{C}))^2 \to \mathbb{R}^4 \times \text{SL}(2, \mathbb{C})$ on $\mathbb{R}^4 \times \text{SL}(2, \mathbb{C})$ by (Fig. 1.5)

Fig. 1.5 Depiction of the spectrum (in red) of $\hat{J}_{R03}$. The spectra of $\hat{J}_{R03}$ and $\hat{J}_{L03}$ coincide

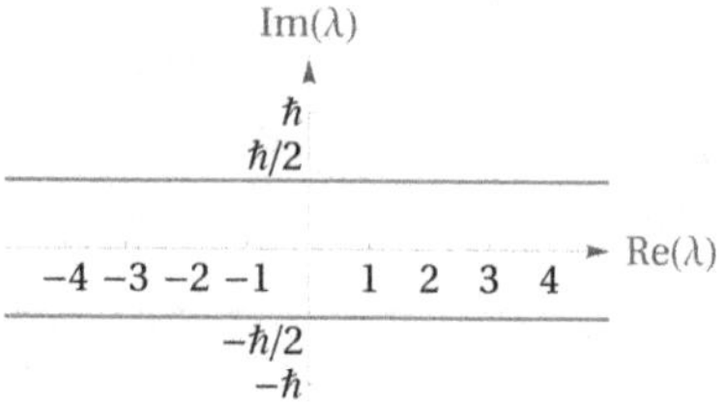

$$(a_1, G_1) \cdot (a_2, G_2) := (a_1 + \Phi_2(G_1)a_2, G_1 \cdot G_2) \ ,$$

for all $(a_1, G_1), (a_2, G_2) \in \mathbb{R}^4 \times \mathrm{SL}(2, \mathbb{C})$. Then, it follows for $(a, G), (a_1, G_1), (a_2, G_2), (a_3, G_3) \in \mathbb{R}^4 \times \mathrm{SL}(2, \mathbb{C})$ that

$$\begin{aligned}
&(a_1, G_1) \cdot ((a_2, G_2) \cdot (a_3, G_3)) = (a_1, G_1) \cdot (a_2 + \Phi_2(G_2)a_3, G_2 \cdot G_3) \\
&= (a_1 + \Phi_2(G_1)(a_2 + \Phi_2(G_2)a_3), G_1 \cdot (G_2 \cdot G_3)) \\
&= (a_1 + \Phi_2(G_1)a_2 + \Phi_2(G_1 \cdot G_2)a_3, (G_1 \cdot G_2) \cdot G_3) \\
&= (a_1 + \Phi_2(G_1)a_2, G_1 \cdot G_2) \cdot (a_3, G_3) \\
&= ((a_1, G_1) \cdot (a_2, G_2)) \cdot (a_3, G_3) \ , \\
&(a, G) \cdot (0, E) = (a + \Phi_2(G) \cdot 0, G \cdot E) = (a, G) \ , \\
&(0, E) \cdot (a, G) = (0 + \Phi_2(E) \cdot a, E \cdot G) = (a, G) \ , \\
&(a, G) \cdot (-\Phi_2(G^{-1})\, a, G^{-1}) = (a + \Phi_2(G)(-[\Phi_2(G)]^{-1}a), G \cdot G^{-1}) = (0, E) \ , \\
&(-\Phi_2(G^{-1})\, a, G^{-1}) \cdot (a, G) = (-\Phi_2(G^{-1})\, a + \Phi_2(G^{-1})\, a, G^{-1} \cdot G) = (0, E) \ .
\end{aligned}$$

Hence, $(\mathbb{R}^4 \times \mathrm{SL}(2, \mathbb{C}), \cdot)$ is a group with unit element $(0, E)$, and, for every $(a, G) \in \mathbb{R}^4 \times \mathrm{SL}(2, \mathbb{C})$, the corresponding inverse element is given by $(-\Phi_2(G^{-1})\, a, G^{-1})$.

In the following, we assume the real vector space $\mathbb{R}^4 \times \mathrm{M}(2, \mathbb{C})$ equipped with the norm

$$\|(a, M)\| := (\,|a|^2 + \|M\|_{\mathrm{op}}^2)^{1/2} \ ,$$

for every $(a, M) \in \mathbb{R}^4 \times \mathrm{M}(2, \mathbb{C})$, where $|\ |$ denotes the Euclidean norm on $\mathbb{R}^4$. Since $(\mathbb{R}^4, |\ |)$ and $(\mathrm{M}(2, \mathbb{C}), \|\ \|_{\mathrm{op}})$ are complete, $(\mathbb{R}^4 \times \mathrm{M}(2, \mathbb{C}), \|\ \|)$ is a Banach space, see e.g. Lemma 12.2.7 in the Appendix of [7]. In particular, the coordinate projections onto $\mathbb{R}^4$ and $\mathrm{M}(2, \mathbb{C})$ are continuous. In addition to $\|\ \|$, we are also going to use the equivalent maximum norm $\| \ \|_\infty$ on $\mathbb{R}^4 \times \mathrm{M}(2, \mathbb{C})$, where, for $(a, M) \in \mathbb{R}^4 \times \mathrm{M}(2, \mathbb{C})$, $\|(a, M)\|_\infty$ is defined as the maximum of the absolute values of the components of a and M. That the norms $\| \ \|$ and $\| \ \|_\infty$ are equivalent can be seen as follows.[3] For $(a, M) \in \mathbb{R}^4 \times \mathrm{M}(2, \mathbb{C})$, we have that

$$|a_k| \leqslant |a| \ , \ \ |M_{jk}| \leqslant \|M\|_\infty \leqslant 2^{1/2} \, \|M\|_{\mathrm{op}} \ ,$$

for $j, k \in \{1, \dots, 4\}$, where we used Inequality 3.1 from Part 1, and hence that

$$\|(a, M)\|_\infty \leqslant 2^{1/2} \, (|a| + \|M\|_{\mathrm{op}}) \leqslant 2 \, (|a|^2 + \|M\|_{\mathrm{op}}^2)^{1/2} = 2 \, \|(a, M)\| \ .$$

Further,

$$|a| \leqslant \|a\|_\infty \ , \ \ \|M\|_{\mathrm{op}} \leqslant 2^{3/2} \, \|M\|_\infty \ ,$$

where we used Inequality 3.1 from Part 1, and hence

$$|a| + \|M\|_{\mathrm{op}} \leqslant \|a\|_\infty + 2^{3/2} \, \|M\|_\infty \leqslant (1 + 2^{3/2}) \, \|(a, M)\|_\infty \ .$$

The latter implies that

$$\|(a, M)\| = (\,|a|^2 + \|M\|_{\mathrm{op}}^2)^{1/2} \leqslant |a| + \|M\|_{\mathrm{op}} \leqslant (1 + 2^{3/2}) \, \|(a, M)\|_\infty \ .$$

As consequence, we have that

$$\|(a, M)\| \leqslant (1 + 2^{3/2}) \, \|(a, M)\|_\infty \leqslant 2 \, (1 + 2^{3/2}) \, \|(a, M)\| \ ,$$

for every $(a, M) \in \mathbb{R}^4 \times \mathrm{M}(2, \mathbb{C})$, and hence that $\| \ \|$ and $\| \ \|_\infty$ are equivalent.

We equip $\mathbb{R}^4 \times \mathrm{SL}(2, \mathbb{C})$ with the topology induced by that of $(\mathbb{R}^4 \times \mathrm{M}(2, \mathbb{C}), \| \ \|)$. Then $\mathbb{R}^4 \times \mathrm{SL}(2, \mathbb{C})$ is a Hausdorff topological space, with a countable basis.

We note that the multiplication is continuous, since if $((a_{1\nu}, G_{1\nu}))_{\nu \in \mathbb{N}}$ and $((a_{2\nu}, G_{2\nu}))_{\nu \in \mathbb{N}}$ are sequences in $\mathbb{R}^4 \times \mathrm{SL}(2, \mathbb{C})$ that are convergent to (a_1, G_1) and $(a_2, G_2) \in \mathbb{R}^4 \times \mathrm{SL}(2, \mathbb{C})$, respectively, then

[3] This can also be concluded from a general result, see [36], that all norms defined on a finite dimensional vector space are equivalent.

$$
\begin{aligned}
&\|(a_{1\nu}, G_{1\nu}) \cdot (a_{2\nu}, G_{2\nu}) - (a_1, G_1) \cdot (a_2, G_2)\|^2 \\
&= \|(a_{1\nu} + \Phi_2(G_{1\nu})a_{2\nu}, G_{1\nu}G_{2\nu}) - (a_1 + \Phi_2(G_1)a_2, G_1G_2)\|^2 \\
&= \|(a_{1\nu} - a_1 + \Phi_2(G_{1\nu})a_{2\nu} - \Phi_2(G_1)a_2, G_{1\nu}G_{2\nu} - G_1G_2)\|^2 \\
&= |a_{1\nu} - a_1 + \Phi_2(G_{1\nu})a_{2\nu} - \Phi_2(G_1)a_2|^2 + \|G_{1\nu}G_{2\nu} - G_1G_2\|_{\text{op}}^2 \\
&= |a_{1\nu} - a_1 + (\Phi_2(G_{1\nu}) - \Phi_2(G_1))(a_{2\nu} - a_2) + (\Phi_2(G_{1\nu}) - \Phi_2(G_1))a_2 \\
&\quad + \Phi_2(G_1)(a_{2\nu} - a_2)|^2 \\
&\quad + \|(G_{1\nu} - G_1)(G_{2\nu} - G_2) + G_1(G_{2\nu} - G_2) + (G_{1\nu} - G_1)G_2\|_{\text{op}}^2 \\
&\leqslant 4[\,|a_{1\nu} - a_1|^2 + |(\Phi_2(G_{1\nu}) - \Phi_2(G_1))(a_{2\nu} - a_2)|^2 \\
&\quad + |(\Phi_2(G_{1\nu}) - \Phi_2(G_1))a_2|^2 + |\Phi_2(G_1)(a_{2\nu} - a_2)|^2 \\
&\quad + \|(G_{1\nu} - G_1)(G_{2\nu} - G_2)\|_{\text{op}}^2 + \|G_1(G_{2\nu} - G_2)\|_{\text{op}}^2 \\
&\quad + \|(G_{1\nu} - G_1)G_2\|_{\text{op}}^2] \\
&\leqslant 4[\,|a_{1\nu} - a_1|^2 + \|\Phi_2(G_{1\nu}) - \Phi_2(G_1)\|_{\text{op}}^2\,|a_{2\nu} - a_2|^2 \\
&\quad + \|\Phi_2(G_{1\nu}) - \Phi_2(G_1)\|_{\text{op}}^2\,|a_2|^2 + \|\Phi_2(G_1)\|_{\text{op}}^2\,|a_{2\nu} - a_2|^2 \\
&\quad + \|G_{1\nu} - G_1\|_{\text{op}}^2\,\|G_{2\nu} - G_2\|_{\text{op}}^2 + \|G_1\|_{\text{op}}^2\,\|G_{2\nu} - G_2\|_{\text{op}}^2 \\
&\quad + \|G_{1\nu} - G_1\|_{\text{op}}^2\,\|G_2\|_{\text{op}}^2]\ ,
\end{aligned}
$$

for every $\nu \in \mathbb{N}$ and hence

$$
\lim_{\nu\to\infty} (a_{1\nu}, G_{1\nu}) \cdot (a_{2\nu}, G_{2\nu}) = (a_1, G_1) \cdot (a_2, G_2)\ ,
$$

where we used that, according to Corollary 1.4, Φ_2 is continuous. In addition,

$$
\begin{aligned}
&\|(a_{1\nu}, G_{1\nu})^{-1} - (a_1, G_1)^{-1}\|^2 \\
&= \|(-\Phi_2(G_{1\nu}^{-1})a_{1\nu}, G_{1\nu}^{-1}) - (-\Phi_2(G_1^{-1})a_1, G_1^{-1})\|^2 \\
&= |\Phi_2(G_{1\nu}^{-1})a_{1\nu} - \Phi_2(G_1^{-1})a_1|^2 + \|G_{1\nu}^{-1} - G_1^{-1}\|_{\text{op}}^2 \\
&= |(\Phi_2(G_{1\nu}^{-1}) - \Phi_2(G_1^{-1}))(a_{1\nu} - a_1) + \Phi_2(G_1^{-1})(a_{1\nu} - a_1) \\
&\quad + (\Phi_2(G_{1\nu}^{-1}) - \Phi_2(G_1^{-1}))\,a_1|^2 + \|G_{1\nu}^{-1} - G_1^{-1}\|_{\text{op}}^2 \\
&\leqslant 4[\,\|\Phi_2(G_{1\nu}^{-1}) - \Phi_2(G_1^{-1})\|_{\text{op}}^2\,|a_{1\nu} - a_1|^2 + \|\Phi_2(G_1^{-1})\|_{\text{op}}^2\,|a_{1\nu} - a_1|^2 \\
&\quad + \|\Phi_2(G_{1\nu}^{-1}) - \Phi_2(G_1^{-1})\|_{\text{op}}^2\,|a_1|^2\,] + \|G_{1\nu}^{-1} - G_1^{-1}\|_{\text{op}}^2\ ,
\end{aligned}
$$

for every $\nu \in \mathbb{N}$ and hence

$$
\lim_{\nu\to\infty} (a_{1\nu}, G_{1\nu})^{-1} = (a_1, G_1)^{-1}\ ,
$$

where we again used that, according to Corollary 1.4, Φ_2 is continuous.

Therefore, the group operations, i.e., multiplication and inversion, of $\mathbb{R}^4 \times \mathrm{SL}(2,\mathbb{C})$ are continuous, with respect to the induced topology. In this way, $\mathbb{R}^4 \times \mathrm{SL}(2,\mathbb{C})$ becomes a topological group.

Exercise 1.7 Show that $\mathbb{R}^4 \times \mathrm{SL}(2,\mathbb{C}) \subset \mathbb{R}^4 \times \mathrm{M}(2,\mathbb{C})$ is closed and unbounded.

According to Exercise 1.8, by

$$\hat{\Phi}_2(a, G) := (a, \Phi_2(G)) \,, \tag{1.22}$$

for every $(a, G) \in \mathbb{R}^4 \times \mathrm{SL}(2,\mathbb{C})$, there is given a component-wise continuous double cover $\hat{\Phi}_2 : \mathbb{R}^4 \times \mathrm{SL}(2,\mathbb{C}) \to \mathscr{P}_+^\uparrow$ of the restricted Poincaré group, $\mathscr{P}_+^\uparrow$.

Exercise 1.8 Show that by

$$\hat{\Phi}_2(a, G) := (a, \Phi_2(G)) \,,$$

for every $(a, G) \in \mathbb{R}^4 \times \mathrm{SL}(2,\mathbb{C})$, there is given a component-wise continuous epimorphism $\hat{\Phi}_2 : \mathbb{R}^4 \times \mathrm{SL}(2,\mathbb{C}) \to \mathscr{P}_+^\uparrow$. In particular,

$$\hat{\Phi}_2(a_1, G_1) = \hat{\Phi}_2(a_2, G_2) \,,$$

for $(a_1, G_1), (a_2, G_2) \in \mathbb{R}^4 \times \mathrm{SL}(2,\mathbb{C})$, if and only if $a_2 = a_1$ and $G_2 = \pm G_1$ (Fig. 1.6).

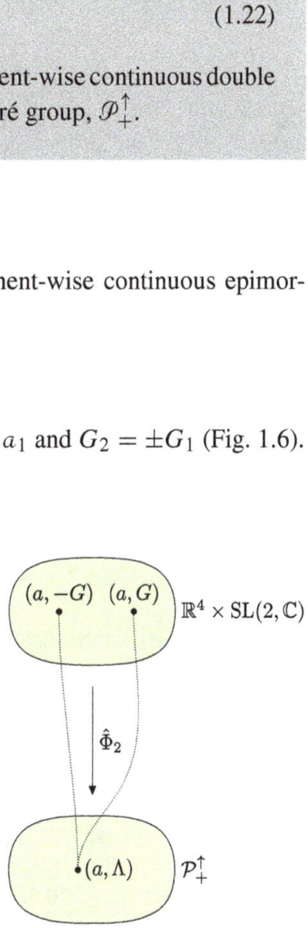

Fig. 1.6 Depiction of the action of $\hat{\Phi}_2$

1.6 An Extension to Strongly Continuous Representations of $\mathbb{R}^4 \times \mathrm{SL}(2, \mathbb{C})$

Again, in the following, let $a \geqslant 0$ and $\hbar \in \{\hbar_R, \hbar_L\}$. Using the strongly continuous unitary representation $\hat{U}_a : \mathcal{P}_+^\uparrow \to L(L^2_{\mathbb{C}}(\mathbb{R}^3, \varphi_a), L^2_{\mathbb{C}}(\mathbb{R}^3, \varphi_a))$, of the restricted Poincaré group $\mathcal{P}_+^\uparrow$ from Theorem 5.10 of Part I, for every $(a, G) \in \mathbb{R}^4 \times \mathrm{SL}(2, \mathbb{C})$, we define $\hat{T}_a^{\hbar}(a, G) : (L^2_{\mathbb{C}}(\mathbb{R}^3, \varphi_a))^2 \to (L^2_{\mathbb{C}}(\mathbb{R}^3, \varphi_a))^2$ by

$$\hat{T}_a^{\hbar}(a, G) \begin{pmatrix} f_1 \\ f_2 \end{pmatrix} := \hbar(G) \cdot \begin{pmatrix} \hat{U}_a(a, \Phi_2(G)) f_1 \\ \hat{U}_a(a, \Phi_2(G)) f_2 \end{pmatrix} \tag{1.23}$$
$$= \begin{pmatrix} [\hbar(G)]_{11} \hat{U}_a(a, \Phi_2(G)) f_1 + [\hbar(G)]_{12} \hat{U}_a(a, \Phi_2(G)) f_2 \\ [\hbar(G)]_{21} \hat{U}_a(a, \Phi_2(G)) f_1 + [\hbar(G)]_{22} \hat{U}_a(a, \Phi_2(G)) f_2 \end{pmatrix} ,$$

for every ${}^t(f_1, f_2) \in (L^2_{\mathbb{C}}(\mathbb{R}^3, \varphi_a))^2$. From the linearity of matrix multiplication and the linearity of the images of $\hat{U}_a$, it follows that $\hat{T}_a^{\hbar}(a, G)$ is linear. Further, since

$$\begin{aligned} \|\hat{T}_a^{\hbar}(a, G) f\|^2 &= \|[\hbar(G)]_{11} \hat{U}_a(a, \Phi_2(G)) f_1 + [\hbar(G)]_{12} \hat{U}_a(a, \Phi_2(G)) f_2\|_2^2 \\ &+ \|[\hbar(G)]_{21} \hat{U}_a(a, \Phi_2(G)) f_1 + [\hbar(G)]_{22} \hat{U}_a(a, \Phi_2(G)) f_2\|_2^2 \\ &\leqslant 2\, \|\hbar(G)\|_\infty^2 \, (\|f_1\|_2 + \|f_2\|_2)^2 \leqslant 4\, \|G\|_\infty^2 \|f\|^2 , \end{aligned}$$

for every $f \in (L^2_{\mathbb{C}}(\mathbb{R}^3, \varphi_a))^2$, where $\langle \, | \, \rangle$ denotes the scalar product for $(L^2_{\mathbb{C}}(\mathbb{R}^3, \varphi_a))^2$, with induced norm $\| \; \|$, and $\langle \, | \, \rangle_2$ denotes the scalar product for $L^2_{\mathbb{C}}(\mathbb{R}^3, \varphi_a)$, with induced norm $\| \; \|_2$, it follows that by (1.23), there is defined a map

$$\hat{T}_a^{\hbar} : \mathbb{R}^4 \times \mathrm{SL}(2, \mathbb{C}) \to L((L^2_{\mathbb{C}}(\mathbb{R}^3, \varphi_a))^2, (L^2_{\mathbb{C}}(\mathbb{R}^3, \varphi_a))^2) .$$

In particular, if $G \in \mathrm{SU}(2)$, we have that $\hbar(G) = G \in \mathrm{SU}(2)$ and therefore that

$$\begin{aligned} &\langle \hat{T}_a^{\hbar}(a, G)\, {}^t(f_1, f_2) | \hat{T}_a^{\hbar}(a, G)\, {}^t(g_1, g_2) \rangle \\ &= \langle [\hbar(G)]_{11} U_a(a, \Phi_2(G)) f_1 + [\hbar(G)]_{12} U_a(a, \Phi_2(G)) f_2 | [\hbar(G)]_{11} U_a(a, \Phi_2(G)) g_1 + [\hbar(G)]_{12} U_a(a, \Phi_2(G)) g_2 \rangle_2 \\ &\quad + \langle [\hbar(G)]_{21} U_a(a, \Phi_2(G)) f_1 + [\hbar(G)]_{22} U_a(a, \Phi_2(G)) f_2 | [\hbar(G)]_{21} U_a(a, \Phi_2(G)) g_1 + [\hbar(G)]_{22} U_a(a, \Phi_2(G)) g_2 \rangle_2 \\ &= |[\hbar(G)]_{11}|^2 \langle f_1 | g_1 \rangle_2 + [\hbar(G)]_{11}^* [\hbar(G)]_{12} \langle f_1 | g_2 \rangle_2 + [\hbar(G)]_{12}^* [\hbar(G)]_{11} \langle f_2 | g_1 \rangle_2 + |[\hbar(G)]_{12}|^2 \langle f_2 | g_2 \rangle_2 \\ &\quad + |[\hbar(G)]_{21}|^2 \langle f_1 | g_1 \rangle_2 + [\hbar(G)]_{21}^* [\hbar(G)]_{22} \langle f_1 | g_2 \rangle_2 + [\hbar(G)]_{22}^* [\hbar(G)]_{21} \langle f_2 | g_1 \rangle_2 + |[\hbar(G)]_{22}|^2 \langle f_2 | g_2 \rangle_2 \\ &= (|[\hbar(G)]_{11}|^2 + |[\hbar(G)]_{21}|^2) \langle f_1 | g_1 \rangle_2 + ([\hbar(G)]_{11}^* [\hbar(G)]_{12} + [\hbar(G)]_{21}^* [\hbar(G)]_{22}) \langle f_1 | g_2 \rangle_2 \\ &\quad + ([\hbar(G)]_{11} [\hbar(G)]_{12}^* + [\hbar(G)]_{21} [\hbar(G)]_{22}^*) \langle f_2 | g_1 \rangle_2 + (|[\hbar(G)]_{12}|^2 + |[\hbar(G)]_{22}|^2) \langle f_2 | g_2 \rangle_2 \\ &= \langle f_1 | g_1 \rangle_2 + \langle f_2 | g_2 \rangle_2 = \langle {}^t(f_1, f_2) | {}^t(g_1, g_2) \rangle , \end{aligned}$$

for all ${}^t(f_1, f_2), {}^t(g_1, g_2) \in (L^2_{\mathbb{C}}(\mathbb{R}^3, \varphi_a))^2$. Hence $\hat{T}^{\hbar}_a(a, G)$ preserves the scalar product.

Further, for $(a_1, G_1), (a_2, G_2) \in \mathbb{R}^4 \times \mathrm{SL}(2, \mathbb{C})$, we have that

$$
\begin{aligned}
&\hat{T}^{\hbar}_a(a_1, G_1)\, \hat{T}^{\hbar}_a(a_2, G_2) f \\
&= \hat{T}^{\hbar}_a(a_1, G_1) \begin{pmatrix} [\hbar(G_2)]_{11} \hat{U}_a(a_2, \Phi_2(G_2)) f_1 + [\hbar(G)]_{12} \hat{U}_a(a_2, \Phi_2(G_2)) f_2 \\ [\hbar(G_2)]_{21} \hat{U}_a(a_2, \Phi_2(G_2)) f_1 + [\hbar(G_2)]_{22} \hat{U}_a(a_2, \Phi_2(G_2)) f_2 \end{pmatrix} \\
&= \hbar(G_1) \cdot \begin{pmatrix} [\hbar(G_2)]_{11} \hat{U}_a(a_1, \Phi_2(G_1)) \hat{U}_a(a_2, \Phi_2(G_2)) f_1 + [\hbar(G)]_{12} \hat{U}_a(a_1, \Phi_2(G_1)) \hat{U}_a(a_2, \Phi_2(G_2)) f_2 \\ [\hbar(G_2)]_{21} \hat{U}_a(a_1, \Phi_2(G_1)) \hat{U}_a(a_2, \Phi_2(G_2)) f_1 + [\hbar(G_2)]_{22} \hat{U}_a(a_1, \Phi_2(G_1)) \hat{U}_a(a_2, \Phi_2(G_2)) f_2 \end{pmatrix} \\
&= \hbar(G_1) \cdot \begin{pmatrix} [\hbar(G_2)]_{11} \hat{U}_a(a_1 + \Phi_2(G_1) a_2, \Phi_2(G_1)\Phi_2(G_2)) f_1 + [\hbar(G)]_{12} \hat{U}_a(a_1 + \Phi_2(G_1) a_2, \Phi_2(G_1)\Phi_2(G_2)) f_2 \\ [\hbar(G_2)]_{21} \hat{U}_a(a_1 + \Phi_2(G_1) a_2, \Phi_2(G_1)\Phi_2(G_2)) f_1 + [\hbar(G_2)]_{22} \hat{U}_a(a_1 + \Phi_2(G_1) a_2, \Phi_2(G_1)\Phi_2(G_2)) f_2 \end{pmatrix} \\
&= \hbar(G_1) \cdot \hbar(G_2) \cdot \begin{pmatrix} \hat{U}_a(a_1 + \Phi_2(G_1) a_2, \Phi_2(G_1 \cdot G_2)) f_1 \\ \hat{U}_a(a_1 + \Phi_2(G_1) a_2, \Phi_2(G_1 \cdot G_2)) f_2 \end{pmatrix} = \hbar(G_1 \cdot G_2) \cdot \begin{pmatrix} \hat{U}_a(a_1 + \Phi_2(G_1) a_2, \Phi_2(G_1 \cdot G_2)) f_1 \\ \hat{U}_a(a_1 + \Phi_2(G_1) a_2, \Phi_2(G_1 \cdot G_2)) f_2 \end{pmatrix} \\
&= \hat{T}^{\hbar}_a((a_1, G_1) \cdot (a_2, G_2)) f \ ,
\end{aligned}
$$

for every $f \in (L^2_{\mathbb{C}}(\mathbb{R}^3, \varphi_a))^2$ and hence that

$$
\hat{T}^{\hbar}_a((a_1, G_1) \cdot (a_2, G_2)) = \hat{T}^{\hbar}_a(a_1, G_1)\, \hat{T}^{\hbar}_a(a_2, G_2) \ .
$$

Since $\hat{T}^{\hbar}_a(0, E)$ is given by the identity map on $(L^2_{\mathbb{C}}(\mathbb{R}^3, \varphi_a))^2$, it follows that $\hat{T}^{\hbar}_a$ is a representation of $\mathbb{R}^4 \times \mathrm{SL}(2, \mathbb{C})$. We note that this implies also that $\hat{T}^{\hbar}_a(a, G)$ is unitary for every $a \in \mathbb{R}^4$ and $G \in \mathrm{SU}(2)$. In this connection, also note Corollary 1.3. Further, if $G, H \in \mathrm{SL}(2, \mathbb{C})$, $a, \mathfrak{b} \in \mathbb{R}^4$ and $f \in (L^2_{\mathbb{C}}(\mathbb{R}^3, \varphi_a))^2$, we have that

$$
\begin{aligned}
&\hat{T}^{\hbar}_a(a, G) f - \hat{T}^{\hbar}_a(\mathfrak{b}, H) f = \hbar(G) \cdot \begin{pmatrix} \hat{U}_a(a, \Phi_2(G)) f_1 \\ \hat{U}_a(a, \Phi_2(G)) f_2 \end{pmatrix} - \hbar(H) \cdot \begin{pmatrix} \hat{U}_a(\mathfrak{b}, \Phi_2(H)) f_1 \\ \hat{U}_a(\mathfrak{b}, \Phi_2(H)) f_2 \end{pmatrix} \\
&= (\hbar(G) - \hbar(H)) \cdot \begin{pmatrix} \hat{U}_a(a, \Phi_2(G)) f_1 \\ \hat{U}_a(a, \Phi_2(G)) f_2 \end{pmatrix} + \hbar(H) \cdot \begin{pmatrix} [\hat{U}_a(a, \Phi_2(G)) - \hat{U}_a(\mathfrak{b}, \Phi_2(H))] f_1 \\ [\hat{U}_a(a, \Phi_2(G)) - \hat{U}_a(\mathfrak{b}, \Phi_2(H))] f_2 \end{pmatrix} \\
&= \begin{pmatrix} ([\hbar(G)]_{11} - [\hbar(H)]_{11}) \hat{U}_a(a, \Phi_2(G)) f_1 + ([\hbar(G)]_{12} - [\hbar(H)]_{12}) \hat{U}_a(a, \Phi_2(G)) f_2 \\ ([\hbar(G)]_{21} - [\hbar(H)]_{21}) \hat{U}_a(a, \Phi_2(G)) f_1 + ([\hbar(G)]_{22} - [\hbar(H)]_{22}) \hat{U}_a(a, \Phi_2(G)) f_2 \end{pmatrix} \\
&\quad + \begin{pmatrix} [\hbar(H)]_{11} [\hat{U}_a(a, \Phi_2(G)) - \hat{U}_a(\mathfrak{b}, \Phi_2(H))] f_1 + [\hbar(H)]_{12} [\hat{U}_a(a, \Phi_2(G)) - \hat{U}_a(\mathfrak{b}, \Phi_2(H))] f_2 \\ [\hbar(H)]_{21} [\hat{U}_a(a, \Phi_2(G)) - \hat{U}_a(\mathfrak{b}, \Phi_2(H))] f_1 + [\hbar(H)]_{22} [\hat{U}_a(a, \Phi_2(G)) - \hat{U}_a(\mathfrak{b}, \Phi_2(H))] f_2 \end{pmatrix}
\end{aligned}
$$

and hence that

$$
\begin{aligned}
&\frac{1}{4}\,\|\hat{T}^{\mathfrak{h}}_a(a,G)f - \hat{T}^{\mathfrak{h}}_a(\mathfrak{b},H)f\|^2 \\
&\leqslant \frac{1}{2}\left\|\begin{pmatrix}([\mathfrak{h}(G)]_{11}-[\mathfrak{h}(H)]_{11})\hat{U}_a(a,\Phi_2(G))f_1+([\mathfrak{h}(G)]_{12}-[\mathfrak{h}(H)]_{12})\hat{U}_a(a,\Phi_2(G))f_2\\ ([\mathfrak{h}(G)]_{21}-[\mathfrak{h}(H)]_{21})\hat{U}_a(a,\Phi_2(G))f_1+([\mathfrak{h}(G)]_{22}-[\mathfrak{h}(H)]_{22})\hat{U}_a(a,\Phi_2(G))f_2\end{pmatrix}\right\|^2 \\
&\quad+\frac{1}{2}\left\|\begin{pmatrix}[\mathfrak{h}(H)]_{11}[\hat{U}_a(a,\Phi_2(G))-\hat{U}_a(\mathfrak{b},\Phi_2(H))]f_1+[\mathfrak{h}(H)]_{12}[\hat{U}_a(a,\Phi_2(G))-\hat{U}_a(\mathfrak{b},\Phi_2(H))]f_2\\ [\mathfrak{h}(H)]_{21}[\hat{U}_a(a,\Phi_2(G))-\hat{U}_a(\mathfrak{b},\Phi_2(H))]f_1+[\mathfrak{h}(H)]_{22}[\hat{U}_a(a,\Phi_2(G))-\hat{U}_a(\mathfrak{b},\Phi_2(H))]f_2\end{pmatrix}\right\|^2 \\
&=\frac{1}{2}\,\|([\mathfrak{h}(G)]_{11}-[\mathfrak{h}(H)]_{11})\hat{U}_a(a,\Phi_2(G))f_1+([\mathfrak{h}(G)]_{12}-[\mathfrak{h}(H)]_{12})\hat{U}_a(a,\Phi_2(G))f_2\|_2^2 \\
&\quad+\frac{1}{2}\,\|([\mathfrak{h}(G)]_{21}-[\mathfrak{h}(H)]_{21})\hat{U}_a(a,\Phi_2(G))f_1+([\mathfrak{h}(G)]_{22}-[\mathfrak{h}(H)]_{22})\hat{U}_a(a,\Phi_2(G))f_2\|_2^2 \\
&\quad+\frac{1}{2}\,\|[\mathfrak{h}(H)]_{11}[\hat{U}_a(a,\Phi_2(G))-\hat{U}_a(\mathfrak{b},\Phi_2(H))]f_1+[\mathfrak{h}(H)]_{12}[\hat{U}_a(a,\Phi_2(G))-\hat{U}_a(\mathfrak{b},\Phi_2(H))]f_2\|_2^2 \\
&\quad+\frac{1}{2}\,\|[\mathfrak{h}(H)]_{21}[\hat{U}_a(a,\Phi_2(G))-\hat{U}_a(\mathfrak{b},\Phi_2(H))]f_1+[\mathfrak{h}(H)]_{22}[\hat{U}_a(a,\Phi_2(G))-\hat{U}_a(\mathfrak{b},\Phi_2(H))]f_2\|_2^2 \\
&\leqslant |[\mathfrak{h}(G)]_{11}-[\mathfrak{h}(H)]_{11}|^2\cdot\|f_1\|^2+|[\mathfrak{h}(G)]_{12}-[\mathfrak{h}(H)]_{12}|^2\cdot\|f_2\|_2^2 \\
&\quad+|[\mathfrak{h}(G)]_{21}-[\mathfrak{h}(H)]_{21}|^2\cdot\|f_1\|_2^2+|[\mathfrak{h}(G)]_{22}-[\mathfrak{h}(H)]_{22}|^2\cdot\|f_2\|_2^2 \\
&\quad+|[\mathfrak{h}(H)]_{11}|^2\cdot\|[\hat{U}_a(a,\Phi_2(G))-\hat{U}_a(\mathfrak{b},\Phi_2(H))]f_1\|_2^2+|[\mathfrak{h}(H)]_{12}|^2\cdot\|[\hat{U}_a(a,\Phi_2(G))-\hat{U}_a(\mathfrak{b},\Phi_2(H))]f_2\|_2^2 \\
&\quad+|[\mathfrak{h}(H)]_{21}|^2\cdot\|[\hat{U}_a(a,\Phi_2(G))-\hat{U}_a(\mathfrak{b},\Phi_2(H))]f_1\|^2+|[\mathfrak{h}(H)]_{22}|^2\cdot\|[\hat{U}_a(a,\Phi_2(G))-\hat{U}_a(\mathfrak{b},\Phi_2(H))]f_2\|_2^2\,.
\end{aligned}
$$

Since $\mathfrak{h}$ and $\hat{\Phi}_2$ are component-wise continuous and $\hat{U}_a$ is strongly continuous, it follows that $\hat{T}^{\mathfrak{h}}_a$ is strongly continuous, i.e., if $(a_1, G_1), (a_2, G_2), \dots$ is a sequence in $\mathbb{R}^4 \times \mathrm{SL}(2, \mathbb{C})$ that converges component-wise to $(a, G) \in \mathbb{R}^4 \times \mathrm{SL}(2, \mathbb{C})$, then

$$\lim_{\nu\to\infty} \|[\hat{T}^{\mathfrak{h}}_a(a_\nu, G_\nu) - \hat{T}^{\mathfrak{h}}_a(a, G)]f\| = 0\,,$$

for every $f \in (L^2_{\mathbb{C}}(\mathbb{R}^3, \varphi_a))^2$.

Hence, we arrive at the following result. If $a \geqslant 0$ and $\mathfrak{h} \in \{\mathfrak{h}_R, \mathfrak{h}_L\}$, then

$$\hat{T}^{\mathfrak{h}}_a : \mathbb{R}^4 \times \mathrm{SL}(2,\mathbb{C}) \to L((L^2_{\mathbb{C}}(\mathbb{R}^3,\varphi_a))^2, (L^2_{\mathbb{C}}(\mathbb{R}^3,\varphi_a))^2)\,,$$

defined by

$$
\begin{aligned}
\hat{T}^{\mathfrak{h}}_a(a,G)\begin{pmatrix}f_1\\ f_2\end{pmatrix} &:= \mathfrak{h}(G)\cdot\begin{pmatrix}\hat{U}_a(a,\Phi_2(G))f_1\\ \hat{U}_a(a,\Phi_2(G))f_2\end{pmatrix} \\
&= \begin{pmatrix}[\mathfrak{h}(G)]_{11}\hat{U}_a(a,\Phi_2(G))f_1+[\mathfrak{h}(G)]_{12}\hat{U}_a(a,\Phi_2(G))f_2\\ [\mathfrak{h}(G)]_{21}\hat{U}_a(a,\Phi_2(G))f_1+[\mathfrak{h}(G)]_{22}\hat{U}_a(a,\Phi_2(G))f_2\end{pmatrix},
\end{aligned}
$$

for every $(a, G) \in \mathbb{R}^4 \times \mathrm{SL}(2, \mathbb{C})$ and ${}^t(f_1, f_2) \in (L^2_{\mathbb{C}}(\mathbb{R}^3, \varphi_a))^2$, is a representation of $\mathbb{R}^4 \times \mathrm{SL}(2, \mathbb{C})$. In addition, $\hat{T}^{\mathfrak{h}}_a$ is strongly continuous, i.e., if $a_1, a_2, \dots$ is a sequence in $\mathbb{R}^4$ that is component-wise convergent to $a \in \mathbb{R}^4$ and $G_1, G_2, \ldots$ is a sequence in $\mathrm{SL}(2, \mathbb{C})$ that converges component-wise to $G \in \mathrm{SL}(2, \mathbb{C})$, then

$$\lim_{\nu \to \infty} \|[\hat{T}_a^{\hbar}(a_\nu, G_\nu) - \hat{T}_a^{\hbar}(a, G)]f\| = 0 ,$$

for every $f \in (L^2_{\mathbb{C}}(\mathbb{R}^3, \varphi_a))^2$. Further, $T_a^{\hbar_R}$ and $T_a^{\hbar_L}$ coincide on $\mathbb{R}^4 \times \mathrm{SU}(2)$ and $T_a^{\hbar_R}(a, G)$ and $T_a^{\hbar_L}(a, G)$ are unitary for every $(a, G) \in \mathbb{R}^4 \times \mathrm{SU}(2)$.

1.6.1 Generators Associated with Translations

For $a \in \mathbb{R}^4$, by $M : \mathbb{R} \to \mathbb{R}^4 \times \mathrm{SL}(2, \mathbb{C})$, defined by

$$M(s) := (sa, E_{2\times 2}) ,$$

for every $s \in \mathbb{R}$, there is defined a one-parameter group, since

$$\begin{aligned} M(0) &= (0, E_{2\times 2}) , \\ M(s_1) \cdot M(s_2) &= (s_1 a, E_{2\times 2}) \cdot (s_2 a, E_{2\times 2}) \\ &= (s_1 a + \Phi_2(E_{2\times 2})\, s_2 a, E_{2\times 2} \cdot E_{2\times 2}) \\ &= ((s_1 + s_2)\, a, E_{2\times 2}) = M(s_1 + s_2) , \end{aligned}$$

for all $s_1, s_2 \in \mathbb{R}$. In addition, M is component-wise continuous. Further, for every $s \in \mathbb{R}$ and $f \in (L^2_{\mathbb{C}}(\mathbb{R}^3, \varphi_a))^2$, we have that

$$\begin{aligned} \hat{T}_a^{\hbar}(M(s)) \begin{pmatrix} f_1 \\ f_2 \end{pmatrix} &= \hbar(E_{2\times 2}) \cdot \begin{pmatrix} \hat{U}_a(sa, \Phi_2(E_{2\times 2})) f_1 \\ \hat{U}_a(sa, \Phi_2(E_{2\times 2})) f_2 \end{pmatrix} \\ &= \begin{pmatrix} \hat{U}_a(sa, E_{4\times 4}) f_1 \\ \hat{U}_a(sa, E_{4\times 4}) f_2 \end{pmatrix} = \begin{pmatrix} \exp(is\,(a_0 v_a^0 - \vec{a} \cdot \mathrm{id}_{\mathbb{R}^3})) f_1 \\ \exp(is\,(a_0 v_a^0 - \vec{a} \cdot \mathrm{id}_{\mathbb{R}^3})) f_2 \end{pmatrix} \\ &= \begin{pmatrix} \exp(is\, T_{a_0 v_a^0 - \vec{a} \cdot \mathrm{id}_{\mathbb{R}^3}}) f_1 \\ \exp(is\, T_{a_0 v_a^0 - \vec{a} \cdot \mathrm{id}_{\mathbb{R}^3}}) f_2 \end{pmatrix} , \end{aligned} \tag{1.24}$$

where $T_{a_0 v_a^0 - \vec{a} \cdot \mathrm{id}_{\mathbb{R}^3}}$ denotes the maximal multiplication operator in $L^2_{\mathbb{C}}(\mathbb{R}^3, \varphi_a)$ corresponding to the function $a_0 v_a^0 - \vec{a} \cdot \mathrm{id}_{\mathbb{R}^3}$, $\vec{a} := {}^t(a_1, a_2, a_3)$ $\vec{a} := {}^t(a_1, a_2, a_3)$, $\vec{a} \cdot \mathrm{id}_{\mathbb{R}^3} : \mathbb{R}^3 \to \mathbb{R}^3$ is defined by $(\vec{a} \cdot \mathrm{id}_{\mathbb{R}^3})(v) := \vec{a} \cdot v$, for every $v \in \mathbb{R}^3$, and we used a result from Part I, Sect. 5.4.1.

Hence, $\hat{T}_a^{\hbar} \circ M$ is a strongly continuous one-parameter unitary group. According to Stone's theorem, there is a unique densely-defined, linear and self-adjoint operator A_M in $X := (L^2_{\mathbb{C}}(\mathbb{R}^3, \varphi_a))^2$ such that

$$\exp(isA_M) = (\hat{T}_a^{\hbar} \circ M)(s) \ ,$$

for every $s \in \mathbb{R}$ and, in particular, that $A_M : D(A_M) \to X$ is given by

$$D(A_M) = \{f \in X : \lim_{s\to 0, s\neq 0} \frac{1}{s}\left[(\hat{T}_a^{\hbar} \circ M)(s) - \mathrm{id}_X\right] f \text{ exists}\}$$

and for every $f \in D(A_M)$

$$A_M f = \frac{1}{i} \lim_{s\to 0, s\neq 0} \frac{1}{s}\left[(\hat{T}_a^{\hbar} \circ M)(s) - \mathrm{id}_X\right] f \ .$$

Further, for $f \in X$ and $s \in \mathbb{R}^*$, it follows that

$$\frac{1}{is}\left[(\hat{T}_a^{\hbar} \circ M)(s) - \mathrm{id}_X\right] f = \begin{pmatrix} \frac{1}{is}\left[\exp(is\, T_{a_0 v_a^0 - \vec{a}\cdot \mathrm{id}_{\mathbb{R}^3}}) - 1\right] f_1 \\ \frac{1}{is}\left[\exp(is\, T_{a_0 v_a^0 - \vec{a}\cdot \mathrm{id}_{\mathbb{R}^3}}) - 1\right] f_2 \end{pmatrix} .$$

We note that the coordinate projections $p_1, p_2 : X \to L^2_{\mathbb{C}}(\mathbb{R}^3, \varphi_a)$ as well as the inclusions $\iota_1, \iota_2 : L^2_{\mathbb{C}}(\mathbb{R}^3, \varphi_a) \hookrightarrow X$, defined by $p_1 f := f_1, p_2 f := f$, for every $f = {}^t(f_1, f_2) \in X$ and $\iota_1 f := {}^t(f, 0)$ and $\iota_2 f := {}^t(0, f)$, for every $f \in L^2_{\mathbb{C}}(\mathbb{R}^3, \varphi_a)$ are linear and continuous, since

$$\|p_k f\|_2 = \|f_k\|_2 \leqslant \|f\| \ , \ \|\iota_k g\| = \|g\|_2 \ ,$$

for every $f = {}^t(f_1, f_2) \in X$, $g \in L^2_{\mathbb{C}}(\mathbb{R}^3, \varphi_a)$ and $k \in \{1, 2\}$. Hence, $f \in D(A_M)$ if and only if f is part of the domain of

$$T_{a_0 v_a^0 - \vec{a}\cdot \mathrm{id}_{\mathbb{R}^3}} \times T_{a_0 v_a^0 - \vec{a}\cdot \mathrm{id}_{\mathbb{R}^3}}$$

and if $f \in D(A_M)$, then

$$A_M f = (T_{a_0 v_a^0 - \vec{a}\cdot \mathrm{id}_{\mathbb{R}^3}} \times T_{a_0 v_a^0 - \vec{a}\cdot \mathrm{id}_{\mathbb{R}^3}}) f \ .$$

Hence, we arrive at the following result.

Generators Associated with Translations in $(L^2_{\mathbb{C}}(\mathbb{R}^3, \varphi_a))^2$ For $a \in \mathbb{R}^4$, we have that

$$(\hat{T}_a^{\hbar} \circ M)(s) = \exp(is\, (T_{a_0 v_a^0 - \vec{a}\cdot \mathrm{id}_{\mathbb{R}^3}} \times T_{a_0 v_a^0 - \vec{a}\cdot \mathrm{id}_{\mathbb{R}^3}})) \ ,$$

for every $s \in \mathbb{R}$, where the one-parameter subgroup of $\mathbb{R}^4 \times \mathrm{SL}(2,\mathbb{C})$, $M : \mathbb{R} \to \mathbb{R}^4 \times \mathrm{SL}(2,\mathbb{C})$ is defined by $M(s) := (sa, E)$, for every $s \in \mathbb{R}$, $T_{a_0 v_a^0 - \vec{a}\cdot \mathrm{id}_{\mathbb{R}^3}}$ denotes the maximal multiplication operator in $L^2_{\mathbb{C}}(\mathbb{R}^3, \varphi_a)$, corresponding to the function $a_0 v_a^0 - \vec{a} \cdot \mathrm{id}_{\mathbb{R}^3}$, $\vec{a} := {}^t(a_1, a_2, a_3)$ and $\vec{a} \cdot \mathrm{id}_{\mathbb{R}^3} : \mathbb{R}^3 \to \mathbb{R}^3$ is defined by $(\vec{a} \cdot \mathrm{id}_{\mathbb{R}^3})(v) := \vec{a} \cdot v$, for every $v \in \mathbb{R}^3$.

Exercise 1.9 Show that the spectrum of $T_{a_0 v_a^0 - \vec{a}\cdot \mathrm{id}_{\mathbb{R}^3}} \times T_{a_0 v_a^0 - \vec{a}\cdot \mathrm{id}_{\mathbb{R}^3}}$ is given by the closure of the range of $a_0 v_a^0 - \vec{a} \cdot \mathrm{id}_{\mathbb{R}^3}$

$$\overline{\mathrm{Ran}(a_0 v_a^0 - \vec{a} \cdot \mathrm{id}_{\mathbb{R}^3})} \ .$$

Exercise 1.10 Show that, if $a \in \mathbb{R}^4$ is future-oriented, timelike, and, in particular, such that $a \cdot a = 1$, the spectrum of $T_{a_0 v_a^0 - \vec{a}\cdot \mathrm{id}_{\mathbb{R}^3}} \times T_{a_0 v_a^0 - \vec{a}\cdot \mathrm{id}_{\mathbb{R}^3}}$ is given by the closed interval $[a, \infty)$.

As a side remark, using the Hilbert space isomorphism $\mathcal{V}_a : L^2_{\mathbb{C}}(\mathbb{R}^3, \varphi_a) \to L^2_{\mathbb{C}}(\mathbb{R}^3)$ from Exercise 5.17 of Part I, defined by

$$\mathcal{V}_a f := (v_a^0)^{-1/2} f \ ,$$

for every $f \in L^2_{\mathbb{C}}(\mathbb{R}^3, \varphi_a)$, with inverse $\mathcal{V}_a^{-1} : L^2_{\mathbb{C}}(\mathbb{R}^3) \to L^2_{\mathbb{C}}(\mathbb{R}^3, \varphi_a)$, given by

$$\mathcal{V}_a^{-1} f := (v_a^0)^{1/2} f \ ,$$

for every $f \in L^2_{\mathbb{C}}(\mathbb{R}^3)$, we note that

$$\mathcal{V}_a \times \mathcal{V}_a : (L^2_{\mathbb{C}}(\mathbb{R}^3, \varphi_a))^2 \to (L^2_{\mathbb{C}}(\mathbb{R}^3))^2$$

is a Hilbert space isomorphism and that

$$(\mathcal{V}_a \times \mathcal{V}_a)\, \hat{T}_a^{\hbar}(M(s))\, (\mathcal{V}_a \times \mathcal{V}_a)^{-1} f = \exp(is\, (T_{a_0 v_a^0 - \vec{a}\cdot \mathrm{id}_{\mathbb{R}^3}} \times T_{a_0 v_a^0 - \vec{a}\cdot \mathrm{id}_{\mathbb{R}^3}})) f \ ,$$

for every $s \in \mathbb{R}$ and $f \in (L^2_{\mathbb{C}}(\mathbb{R}^3))^2$, where $T_{a_0 v_a^0 - \vec{a}\cdot \mathrm{id}_{\mathbb{R}^3}}$ denotes the maximal multiplication operator in $L^2_{\mathbb{C}}(\mathbb{R}^3)$, corresponding to the function $a_0 v_a^0 - \vec{a} \cdot \mathrm{id}_{\mathbb{R}^3}$, $\vec{a} := {}^t(a_1, a_2, a_3)$. As a consequence, we also have the following.

Generators Associated with Translations in $(L^2_{\mathbb{C}}(\mathbb{R}^3))^2$ For $a \in \mathbb{R}^4$, we have that

$$(\mathcal{V}_a \times \mathcal{V}_a)\, \hat{T}_a^{\hbar}(M(s))\, (\mathcal{V}_a \times \mathcal{V}_a)^{-1} = \exp(is\, (T_{a_0 v_a^0 - \vec{a}\cdot \mathrm{id}_{\mathbb{R}^3}} \times T_{a_0 v_a^0 - \vec{a}\cdot \mathrm{id}_{\mathbb{R}^3}})) \ ,$$

for every $s \in \mathbb{R}$, where the one-parameter subgroup of $\mathbb{R}^4 \times \mathrm{SL}(2,\mathbb{C})$, $M : \mathbb{R} \to \mathbb{R}^4 \times \mathrm{SL}(2,\mathbb{C})$ is defined by $M(s) := (sa, E)$, for every $s \in \mathbb{R}$, $T_{a_0 v_a^0 - \vec{a}\cdot \mathrm{id}_{\mathbb{R}^3}}$ denotes the maximal multiplication operator in $L^2_{\mathbb{C}}(\mathbb{R}^3)$, corresponding to the function $a_0 v_a^0 - \vec{a} \cdot \mathrm{id}_{\mathbb{R}^3}$, $\vec{a} := {}^t(a_1, a_2, a_3)$ and $\vec{a} \cdot \mathrm{id}_{\mathbb{R}^3} : \mathbb{R}^3 \to \mathbb{R}^3$ is defined by $(\vec{a} \cdot \mathrm{id}_{\mathbb{R}^3})(v) := \vec{a} \cdot v$, for every $v \in \mathbb{R}^3$.

1.7 Weyl Equations

For $f \in (C^2(\mathbb{R}^4, \mathbb{C}))^2$, we have that

$$\sum_{l=1}^{3} \sigma_l \frac{\partial}{\partial u_l} \begin{pmatrix} f_1 \\ f_2 \end{pmatrix} = \sum_{l=1}^{3} \sigma_l \cdot \begin{pmatrix} \frac{\partial f_1}{\partial u_l} \\ \frac{\partial f_2}{\partial u_l} \end{pmatrix} = \sum_{l=1}^{3} \begin{pmatrix} \sum_{\beta=1}^{2} (\sigma_l)_{1\beta} \frac{\partial f_\beta}{\partial u_l} \\ \sum_{\beta=1}^{2} (\sigma_l)_{2\beta} \frac{\partial f_\beta}{\partial u_l} \end{pmatrix}$$
$$= \begin{pmatrix} \sum_{l=1}^{3} \sum_{\beta=1}^{2} (\sigma_l)_{1\beta} \frac{\partial f_\beta}{\partial u_l} \\ \sum_{l=1}^{3} \sum_{\beta=1}^{2} (\sigma_l)_{2\beta} \frac{\partial f_\beta}{\partial u_l} \end{pmatrix}$$

and further that

$$\sum_{k=1}^{3} \sigma_k \frac{\partial}{\partial u_k} \sum_{l=1}^{3} \sigma_l \frac{\partial}{\partial u_l} \begin{pmatrix} f_1 \\ f_2 \end{pmatrix} = \sum_{k=1}^{3} \sigma_k \cdot \begin{pmatrix} \sum_{l=1}^{3} \sum_{\beta=1}^{2} (\sigma_l)_{1\beta} \frac{\partial^2 f_\beta}{\partial u_k \partial u_l} \\ \sum_{l=1}^{3} \sum_{\beta=1}^{2} (\sigma_l)_{2\beta} \frac{\partial^2 f_\beta}{\partial u_k \partial u_l} \end{pmatrix}$$
$$= \sum_{k=1}^{3} \begin{pmatrix} \sum_{l=1}^{3} \sum_{\alpha=1}^{2} \sum_{\beta=1}^{2} (\sigma_k)_{1\alpha} (\sigma_l)_{\alpha\beta} \frac{\partial^2 f_\beta}{\partial u_k \partial u_l} \\ \sum_{l=1}^{3} \sum_{\alpha=1}^{2} \sum_{\beta=1}^{2} (\sigma_k)_{2\alpha} (\sigma_l)_{\alpha\beta} \frac{\partial^2 f_\beta}{\partial u_k \partial u_l} \end{pmatrix}$$
$$= \begin{pmatrix} \sum_{k=1}^{3} \sum_{l=1}^{3} \sum_{\alpha=1}^{2} \sum_{\beta=1}^{2} (\sigma_k)_{1\alpha} (\sigma_l)_{\alpha\beta} \frac{\partial^2 f_\beta}{\partial u_k \partial u_l} \\ \sum_{k=1}^{3} \sum_{l=1}^{3} \sum_{\alpha=1}^{2} \sum_{\beta=1}^{2} (\sigma_k)_{2\alpha} (\sigma_l)_{\alpha\beta} \frac{\partial^2 f_\beta}{\partial u_k \partial u_l} \end{pmatrix}$$
$$= \begin{pmatrix} \sum_{k=1}^{3} \sum_{l=1}^{3} \sum_{\beta=1}^{2} (\sigma_k \cdot \sigma_l)_{1\beta} \frac{\partial^2 f_\beta}{\partial u_k \partial u_l} \\ \sum_{k=1}^{3} \sum_{l=1}^{3} \sum_{\beta=1}^{2} (\sigma_k \cdot \sigma_l)_{2\beta} \frac{\partial^2 f_\beta}{\partial u_k \partial u_l} \end{pmatrix},$$
$$= \frac{1}{2} \begin{pmatrix} \sum_{k=1}^{3} \sum_{l=1}^{3} \sum_{\beta=1}^{2} (\sigma_k \cdot \sigma_l + \sigma_l \cdot \sigma_k)_{1\beta} \frac{\partial^2 f_\beta}{\partial u_k \partial u_l} \\ \sum_{k=1}^{3} \sum_{l=1}^{3} \sum_{\beta=1}^{2} (\sigma_k \cdot \sigma_l + \sigma_l \cdot \sigma_k)_{2\beta} \frac{\partial^2 f_\beta}{\partial u_k \partial u_l} \end{pmatrix}$$
$$= \frac{1}{2} \begin{pmatrix} \sum_{k=1}^{3} \sum_{l=1}^{3} \sum_{\beta=1}^{2} (2\delta_{kl} E)_{1\beta} \frac{\partial^2 f_\beta}{\partial u_k \partial u_l} \\ \sum_{k=1}^{3} \sum_{l=1}^{3} \sum_{\beta=1}^{2} (2\delta_{kl} E)_{2\beta} \frac{\partial^2 f_\beta}{\partial u_k \partial u_l} \end{pmatrix}$$
$$= \begin{pmatrix} \sum_{k=1}^{3} \sum_{\beta=1}^{2} \delta_{1\beta} \frac{\partial^2 f_\beta}{\partial u_k^2} \\ \sum_{k=1}^{3} \sum_{\beta=1}^{2} \delta_{2\beta} \frac{\partial^2 f_\beta}{\partial u_k^2} \end{pmatrix} = \begin{pmatrix} \sum_{k=1}^{3} \frac{\partial^2 f_1}{\partial u_k^2} \\ \sum_{k=1}^{3} \frac{\partial^2 f_2}{\partial u_k^2} \end{pmatrix},$$

where $\delta_{\alpha\beta} := 0$, if $\alpha \neq \beta$ and $\delta_{\alpha\beta} := 1$, if $\alpha = \beta$. Hence if $f \in (C^2(\mathbb{R}^4, \mathbb{C}))^2$ is a solution of a Weyl equation, i.e.,

$$\frac{\partial f}{\partial t} = \pm\kappa c \sum_{k=1}^{3} \sigma_k \frac{\partial f}{\partial u_k} = \frac{1}{i\hbar}\left(\mp c \sum_{\alpha=1}^{3} \sigma_\alpha \frac{\hbar\kappa}{i}\frac{\partial f}{\partial u_\alpha}\right), \tag{1.25}$$

then

$$\begin{aligned}\frac{\partial^2 f}{\partial t^2} &= \pm\kappa c \sum_{\alpha=1}^{3} \sigma_\alpha \frac{\partial}{\partial u_\alpha}\frac{\partial f}{\partial t} = (\kappa c)^2 \sum_{\alpha=1}^{3} \sigma_\alpha \frac{\partial}{\partial u_\alpha} \sum_{l=1}^{3} \sigma_l \frac{\partial}{\partial u_l} f \\ &= (\kappa c)^2 \sum_{\alpha=1}^{3} \frac{\partial^2 f}{\partial u_\alpha^2},\end{aligned}$$

where c denotes the speed of light in vacuum, $\kappa > 0$ is a scale factor with dimension 1/length and $t := u_0/(\kappa c)$, i.e., both components of f satisfy the same wave equation [72]. In this sense, the Weyl equations can be regarded as "square roots" of a wave equation, an observation that is attributed to Hermann Weyl, in the context of Dirac's derivation of the equation describing particles with mass $m \geqslant 0$ and Spin 1/2 [16]. This observation indicates that, for the case that the mass of the particle vanishes, i.e., if $m = 0$, for the purpose of obtaining such square roots of wave equations, 2-component spinors suffice, instead of the 4-component spinors employed in Dirac's equation.

There appears to be no a priori reason to expect a connection between Weyl equations and the Weyl representations. However, such connection is provided by the fact that, to every $(a, G) \in \mathbb{R}^4 \times \mathrm{SL}(2, \mathbb{C})$, there corresponds a related symmetry transformation of the Weyl equations, i.e., a transformations that maps solutions of the equations into such solutions. For the demonstration, we denote by $\mathbb{C}^{\mathbb{R}^4}$ the complex vector space of all maps from $\mathbb{R}^4$ to $\mathbb{C}$. Further, for every $(a, G) \in \mathbb{R}^4 \times \mathrm{SL}(2, \mathbb{C})$, we define

$$\begin{aligned}[R_{\hat{h}}((a, G))f](u) &:= \hat{h}(G) \cdot f([\Phi_2(G)]^{-1} \cdot (u - a)) \\ &= \begin{pmatrix} \sum_{\beta=1}^{2} [\hat{h}(G)]_{1\beta} f_\beta([\Phi_2(G)]^{-1} \cdot (u - a)) \\ \sum_{\beta=1}^{2} [\hat{h}(G)]_{2\beta} f_\beta([\Phi_2(G)]^{-1} \cdot (u - a)) \end{pmatrix},\end{aligned}$$

for every $f \in \mathbb{C}^{\mathbb{R}^4} \times \mathbb{C}^{\mathbb{R}^4}$ and $u \in \mathbb{R}^4$. Then, $R_{\hat{h}}((a, G)) \in \mathrm{Hom}(\mathbb{C}^{\mathbb{R}^4} \times \mathbb{C}^{\mathbb{R}^4}, \mathbb{C}^{\mathbb{R}^4} \times \mathbb{C}^{\mathbb{R}^4})$, for every $(a, G) \in \mathbb{R}^4 \times \mathrm{SL}(2, \mathbb{C})$. Further, for $(a_1, G_1), (a_2, G_2) \in \mathbb{R}^4 \times \mathrm{SL}(2, \mathbb{C})$, we have that

$$
\begin{aligned}
&[R_{\hbar}((a_1, G_1) \cdot (a_2, G_2)) f](u) = [R_{\hbar}((a_1 + \Phi_2(G_1) \cdot a_2, G_1 \cdot G_2)) f](u) \\
&= \hbar(G_1 \cdot G_2) \cdot f([\Phi_2(G_1 \cdot G_2)]^{-1} \cdot (u - (a_1 + \Phi_2(G_1) \cdot a_2))) \\
&= \hbar(G_1 \cdot G_2) \cdot f([\Phi_2(G_1 \cdot G_2)]^{-1} \cdot (u - a_1) - [\Phi_2(G_2)]^{-1} \cdot a_2) , \\
&[R_{\hbar}((a_1, G_1))[R_{\hbar}((a_2, G_2)) f]](u) \\
&= \hbar(G_1) \cdot [R_{\hbar}((a_2, G_2)) f]([\Phi_2(G_1)]^{-1} \cdot (u - a_1)) \\
&= \hbar(G_1) \cdot \hbar(G_2) \cdot f([\Phi_2(G_2)]^{-1} \cdot ([\Phi_2(G_1)]^{-1} \cdot (u - a_1) - a_2)) \\
&= \hbar(G_1 \cdot G_2) \cdot f([\Phi_2(G_1 \cdot G_2)]^{-1} \cdot (u - a_1) - [\Phi_2(G_2)]^{-1} \cdot a_2))
\end{aligned}
$$

and that

$$
[R_{\hbar}((0, E)) f](u) = \hbar(E) \cdot f([\Phi_2(E)]^{-1} \cdot u) = f(u) ,
$$

for every $f \in \mathbb{C}^{\mathbb{R}^4} \times \mathbb{C}^{\mathbb{R}^4}$ and $u \in \mathbb{R}^4$. Hence,

$$
\begin{aligned}
&R_{\hbar}((a_1, G_1) \cdot (a_2, G_2)) = R_{\hbar}((a_1, G_1)) \circ R_{\hbar}((a_2, G_2)) , \\
&R_{\hbar}((0, E)) = \mathrm{id}_{\mathbb{C}^{\mathbb{R}^4} \times \mathbb{C}^{\mathbb{R}^4}} .
\end{aligned}
$$

As a consequence, the map $R_{\hbar} : \mathbb{R}^4 \times \mathrm{SL}(2, \mathbb{C}) \to \mathrm{Hom}(\mathbb{C}^{\mathbb{R}^4} \times \mathbb{C}^{\mathbb{R}^4}, \mathbb{C}^{\mathbb{R}^4} \times \mathbb{C}^{\mathbb{R}^4})$, that associates with every $(a, G) \in \mathbb{R}^4 \times \mathrm{SL}(2, \mathbb{C})$ the map $R_{\hbar}((a, G))$, is a representation of $\mathbb{R}^4 \times \mathrm{SL}(2, \mathbb{C})$. Further, $R_{\hbar}$ leaves $(C^1(\mathbb{R}^4, \mathbb{C}))^2$ invariant, and for $(a, G) \in \mathbb{R}^4 \times \mathrm{SL}(2, \mathbb{C})$ and $f \in (C^1(\mathbb{R}^4, \mathbb{C}))^2$, we have that

$$
\begin{aligned}
&\frac{\partial R_{\hbar}((a, G)) f}{\partial u_k}(u) \\
&= \begin{pmatrix} \sum_{\beta=1}^{2} [\hbar(G)]_{1\beta} f_{\beta}'([\Phi_2(G)]^{-1} \cdot (u - a)) \cdot [\Phi_2(G)]^{-1} \cdot e_k \\ \sum_{\beta=1}^{2} [\hbar(G)]_{2\beta} f_{\beta}'([\Phi_2(G)]^{-1} \cdot (u - a)) \cdot [\Phi_2(G)]^{-1} \cdot e_k \end{pmatrix} \\
&= \begin{pmatrix} \sum_{\beta=1}^{2} [\hbar(G)]_{1\beta} \sum_{l=0}^{3} \frac{\partial f_{\beta}}{\partial u_l}([\Phi_2(G)]^{-1} \cdot (u - a)) \cdot ([\Phi_2(G)]^{-1})_{lk} \\ \sum_{\beta=1}^{2} [\hbar(G)]_{2\beta} \sum_{l=0}^{3} \frac{\partial f_{\beta}}{\partial u_l}([\Phi_2(G)]^{-1} \cdot (u - a)) \cdot ([\Phi_2(G)]^{-1})_{lk} \end{pmatrix} \\
&= \hbar(G) \cdot \begin{pmatrix} \sum_{l=0}^{3} \frac{\partial f_1}{\partial u_l}([\Phi_2(G)]^{-1} \cdot (u - a)) \, ([\Phi_2(G)]^{-1})_{lk} \\ \sum_{l=0}^{3} \frac{\partial f_2}{\partial u_l}([\Phi_2(G)]^{-1} \cdot (u - a)) \, ([\Phi_2(G)]^{-1})_{lk} \end{pmatrix} \\
&= \sum_{l=0}^{3} ([\Phi_2(G)]^{-1})_{lk} \, \hbar(G) \cdot \frac{\partial f}{\partial u_l}([\Phi_2(G)]^{-1} \cdot (u - a)) ,
\end{aligned}
$$

for every $u \in \mathbb{R}^4$. Hence, we have that

$$
\begin{aligned}
&\left(\sum_{k=0}^{3}\sigma_k\cdot\frac{\partial R_{\hbar_R}((a,G))f}{\partial u_k}\right)(u)\\
&=\sum_{k=0}^{3}\sigma_k\cdot\sum_{l=0}^{3}([\Phi_2(G)]^{-1})_{lk}\ \hbar_R(G)\cdot\frac{\partial f}{\partial u_l}([\Phi_2(G)]^{-1}\cdot(u-a))\\
&=\sum_{l=0}^{3}\left[\sum_{k=0}^{3}([\Phi_2(G)]^{-1})_{lk}\ \sigma_k\right]\cdot\hbar_R(G)\cdot\frac{\partial f}{\partial u_l}([\Phi_2(G)]^{-1}\cdot(u-a))\\
&=\sum_{l=0}^{3}(G^{-1})^*\cdot\sigma_l\cdot G^{-1}\cdot G\cdot\frac{\partial f}{\partial u_l}([\Phi_2(G)]^{-1}\cdot(u-a))\\
&=(G^{-1})^*\cdot\sum_{l=0}^{3}\sigma_l\cdot\frac{\partial f}{\partial u_l}([\Phi_2(G)]^{-1}\cdot(u-a))\\
&=(G^*)^{-1}\cdot\sum_{l=0}^{3}\sigma_l\cdot\frac{\partial f}{\partial u_l}([\Phi_2(G)]^{-1}\cdot(u-a))\\
&=\hbar_L(G)\cdot\left(\sum_{l=0}^{3}\sigma_l\cdot\frac{\partial f}{\partial u_l}\right)([\Phi_2(G)]^{-1}\cdot(u-a))\ ,
\end{aligned}
$$

for every $u\in\mathbb{R}^4$, where we used (1.13), and

$$
\begin{aligned}
&\left(\sum_{k=0}^{3}\bar{\sigma}_k\cdot\frac{\partial R_{\hbar_L}((a,G))f}{\partial u_k}\right)(u)\\
&=\sum_{k=0}^{3}\bar{\sigma}_k\cdot\sum_{l=0}^{3}([\Phi_2(G)]^{-1})_{lk}\ \hbar_L(G)\cdot\frac{\partial f}{\partial u_l}([\Phi_2(G)]^{-1}\cdot(u-a))\\
&=\sum_{l=0}^{3}\left[\sum_{k=0}^{3}([\Phi_2(G)]^{-1})_{lk}\ \bar{\sigma}_k\right]\cdot\hbar_L(G)\cdot\frac{\partial f}{\partial u_l}([\Phi_2(G)]^{-1}\cdot(u-a))\\
&=\left[\sum_{k=0}^{3}([\Phi_2(G)]^{-1})_{0k}\ \bar{\sigma}_k\right]\cdot\hbar_L(G)\cdot\frac{\partial f}{\partial u_0}([\Phi_2(G)]^{-1}\cdot(u-a))\\
&\quad+\sum_{\alpha=0}^{3}\left[\sum_{k=0}^{3}([\Phi_2(G)]^{-1}_{\alpha k}\ \bar{\sigma}_k\right]\cdot\hbar_L(G)\cdot\frac{\partial f}{\partial u_\alpha}([\Phi_2(G)]^{-1}\cdot(u-a))\\
&=\left[[\Phi_2(G)]_{00}\ \bar{\sigma}_0-\sum_{\beta=1}^{3}[\Phi_2(G)]_{\beta 0}\ \bar{\sigma}_\beta\right]\cdot\hbar_L(G)\cdot\frac{\partial f}{\partial u_0}([\Phi_2(G)]^{-1}\cdot(u-a))\\
&\quad-\sum_{\alpha=0}^{3}\left[[\Phi_2(G)]_{0\alpha}\ \bar{\sigma}_0-\sum_{\beta=1}^{3}[\Phi_2(G)]_{\beta\alpha}\ \bar{\sigma}_\beta\right]\cdot\hbar_L(G)
\end{aligned}
$$

$$
\begin{aligned}
&\cdot \frac{\partial f}{\partial u_\alpha}([\Phi_2(G)]^{-1} \cdot (u-a)) \\
&= \left[\sum_{k=0}^{3} [\Phi_2(G)]_{k0}\, \sigma_k \right] \cdot \hbar_L(G) \cdot \frac{\partial f}{\partial u_0}([\Phi_2(G)]^{-1} \cdot (u-a)) \\
&\quad - \sum_{\alpha=0}^{3} \left[\sum_{k=0}^{3} [\Phi_2(G)]_{k\alpha}\, \sigma_k \right] \cdot \hbar_L(G) \cdot \frac{\partial f}{\partial u_\alpha}([\Phi_2(G)]^{-1} \cdot (u-a)) \\
&= G \cdot \sigma_0 \cdot G^* \cdot \hbar_L(G) \cdot \frac{\partial f}{\partial u_0}([\Phi_2(G)]^{-1} \cdot (u-a)) \\
&\quad - \sum_{\alpha=0}^{3} \left(G \cdot \sigma_\alpha \cdot G^* \right) \cdot \hbar_L(G) \cdot \frac{\partial f}{\partial u_\alpha}([\Phi_2(G)]^{-1} \cdot (u-a)) \\
&= G \cdot \sigma_0 \cdot G^* \cdot (G^*)^{-1} \cdot \frac{\partial f}{\partial u_0}([\Phi_2(G)]^{-1} \cdot (u-a)) \\
&\quad - \sum_{\alpha=0}^{3} \left(G \cdot \sigma_\alpha \cdot G^* \right) \cdot (G^*)^{-1} \cdot \frac{\partial f}{\partial u_\alpha}([\Phi_2(G)]^{-1} \cdot (u-a)) \\
&= G \cdot \bar{\sigma}_0 \cdot \frac{\partial f}{\partial u_0}([\Phi_2(G)]^{-1} \cdot (u-a)) + \sum_{\alpha=0}^{3} G \cdot \bar{\sigma}_\alpha \cdot \frac{\partial f}{\partial u_\alpha}([\Phi_2(G)]^{-1} \cdot (u-a)) \\
&= G \cdot \sum_{k=0}^{3} \bar{\sigma}_k \cdot \frac{\partial f}{\partial u_k}([\Phi_2(G)]^{-1} \cdot (u-a)) \\
&= \hbar_R(G) \cdot \left(\sum_{k=0}^{3} \bar{\sigma}_k \cdot \frac{\partial f}{\partial u_k} \right)([\Phi_2(G)]^{-1} \cdot (u-a)) \ ,
\end{aligned}
$$

for every $u \in \mathbb{R}^4$, where

$$
\bar{\sigma}_0 := \sigma_0 \ , \ \bar{\sigma}_1 := -\sigma_1 \ , \ \bar{\sigma}_2 := -\sigma_2 \ , \ \bar{\sigma}_3 := -\sigma_3 \ ,
$$

and we used (1.11) and Formula 5.4 of Part I.

Summarizing the previous, for $f \in (C^1(\mathbb{R}^4, \mathbb{C}))^2$ and $(a, G) \in \mathbb{R}^4 \times \mathrm{SL}(2, \mathbb{C})$, we have the following transformation properties

$$
\sum_{k=0}^{3} \sigma_k \cdot \frac{\partial R_{\hbar_R}((a, G)) f}{\partial u_k} = [R_{\hbar_L}((a, G))]\Big(\sum_{k=0}^{3} \sigma_k \cdot \frac{\partial f}{\partial u_l} \Big) \ ,
$$

$$
\sum_{k=0}^{3} \bar{\sigma}_k \cdot \frac{\partial R_{\hbar_L}((a, G)) f}{\partial u_k} = [R_{\hbar_R}((a, G))]\Big(\sum_{k=0}^{3} \bar{\sigma}_k \cdot \frac{\partial f}{\partial u_l} \Big) \ ,
$$

where

$$\sigma_0 := \begin{pmatrix} 1 & 0 \\ 0 & 1 \end{pmatrix} , \ \sigma_1 := \begin{pmatrix} 0 & 1 \\ 1 & 0 \end{pmatrix} , \ \sigma_2 := \begin{pmatrix} 0 & -i \\ i & 0 \end{pmatrix} , \ \sigma_3 := \begin{pmatrix} 1 & 0 \\ 0 & -1 \end{pmatrix}$$

are the Pauli spin matrices,

$$\bar{\sigma}_0 := \sigma_0 \ , \ \bar{\sigma}_1 := -\sigma_1 \ , \ \bar{\sigma}_2 := -\sigma_2 \ , \ \bar{\sigma}_3 := -\sigma_3 \ ,$$

and for $\hbar \in \{\hbar_R, \hbar_L\}$

$$[R_\hbar((a, G)) f](u) := \hbar(G) \cdot f([\Phi_2(G)]^{-1} \cdot (u - a)) \ ,$$

for every $u \in \mathbb{R}^4$.

As a consequence, if $f \in (C^1(\mathbb{R}^4, \mathbb{C}))^2$ is a solution of a Weyl equation, where the projection $t : \mathbb{R}^4 \to \mathbb{R}$ is defined by $t := (\kappa c)^{-1} u_0$, then, for every $(a, G) \in \mathbb{R}^4 \times \mathrm{SL}(2, \mathbb{C})$, $R_{\hbar_R}((a, G)) f$ resp. $R_{\hbar_L}((a, G)) f$ is a solution of the same equation.

In the following, we represent the Weyl equations in form of Schrödinger equations in $(L^2_{\mathbb{C}}(\mathbb{R}^3, \varphi_a))^2$, where $a \geqslant 0$. For this purpose, we consider the linear operator

$$A_W : (D(T_{v_1}) \cap D(T_{v_2}) \cap D(T_{v_3}))^2 \to (L^2_{\mathbb{C}}(\mathbb{R}^3, \varphi_a))^2 \ ,$$

defined by

$$\begin{aligned} A_W f &:= \sum_{\alpha=1}^{3} \sigma_\alpha \cdot \begin{pmatrix} T_{v_\alpha} f_1 \\ T_{v_\alpha} f_2 \end{pmatrix} \\ &= \begin{pmatrix} T_{v_1} f_2 - i\, T_{v_2} f_2 + T_{v_3} f_1 \\ T_{v_1} f_1 + i\, T_{v_2} f_1 - T_{v_3} f_2 \end{pmatrix} = \begin{pmatrix} v_3 & v_1 - i\, v_2 \\ v_1 + i\, v_2 & -v_3 \end{pmatrix} \cdot \begin{pmatrix} f_1 \\ f_2 \end{pmatrix} \end{aligned}$$

for every $f \in (D(T_{v_1}) \cap D(T_{v_2}) \cap D(T_{v_3}))^2$, where, for $\alpha \in \{1, 2, 3\}$, $v_\alpha : \mathbb{R}^3 \to \mathbb{R}$ is the coordinate projection of $\mathbb{R}^3$ onto the α-th component and T_{v_α} denotes the maximal multiplication operator in $L^2_{\mathbb{C}}(\mathbb{R}^3, \varphi_a)$ with v_α, with domain $D(T_{v_\alpha})$. Associated with the Weyl equations are the Hamilton operators

$$H_L = -\hbar\kappa c A_W \ , \ H_R = \hbar\kappa c A_W \ .$$

We note that A_W is densely-defined, since $(C_0(\mathbb{R}^3, \mathbb{C}))^2 \subset (D(T_{v_1}) \cap D(T_{v_2}) \cap D(T_{v_3}))^2$. Further, A_W is symmetric, since we have that

$$\begin{aligned}
\langle f|A_W g\rangle &= \langle f|\sum_{\alpha=1}^{3}\sigma_\alpha \cdot \begin{pmatrix} T_{v_\alpha} g_1 \\ T_{v_\alpha} g_2 \end{pmatrix}\rangle \\
&= \langle f_1|T_{v_1} g_2 - i\,T_{v_2} g_2 + T_{v_3} g_1\rangle_2 + \langle f_2|T_{v_1} g_1 + i\,T_{v_2} g_1 - T_{v_3} g_2\rangle_2 \\
&= \langle f_1|T_{v_1} g_2\rangle_2 + \langle f_1| - i\,T_{v_2} g_2\rangle_2 + \langle f_1|T_{v_3} g_1\rangle_2 + \langle f_2|T_{v_1} g_1\rangle_2 + \langle f_2|i\,T_{v_2} g_1\rangle_2 + \langle f_2| - T_{v_3} g_2\rangle_2 \\
&= \langle T_{v_1} f_1|g_2\rangle_2 + \langle i\,T_{v_2} f_1|g_2\rangle_2 + \langle T_{v_3} f_1|g_1\rangle_2 + \langle T_{v_1} f_2|g_1\rangle_2 + \langle -i\,T_{v_2} f_2|g_1\rangle_2 + \langle -T_{v_3} f_2|g_2\rangle_2 \\
&= \langle T_{v_3} f_1|g_1\rangle_2 + \langle T_{v_1} f_2|g_1\rangle_2 + \langle -i\,T_{v_2} f_2|g_1\rangle_2 + \langle T_{v_1} f_1|g_2\rangle_2 + \langle i\,T_{v_2} f_1|g_2\rangle_2 + \langle -T_{v_3} f_2|g_2\rangle_2 \\
&= \langle T_{v_3} f_1 + T_{v_1} f_2 - i\,T_{v_2} f_2|g_1\rangle_2 + \langle T_{v_1} f_1 + i\,T_{v_2} f_1 - T_{v_3} f_2|g_2\rangle_2 \\
&= \langle T_{v_1} f_2 - i\,T_{v_2} f_2 + T_{v_3} f_1|g_1\rangle_2 + \langle T_{v_1} f_1 + i\,T_{v_2} f_1 - T_{v_3} f_2|g_2\rangle_2 \\
&= \langle \sum_{\alpha=1}^{3}\sigma_\alpha \cdot \begin{pmatrix} T_{v_\alpha} f_1 \\ T_{v_\alpha} f_2 \end{pmatrix}|g\rangle = \langle A_W f|g\rangle \ ,
\end{aligned}$$

for all $f, g \in (D(T_{v_1}) \cap D(T_{v_2}) \cap D(T_{v_3}))^2$, where $\langle\,|\,\rangle$ denotes the scalar product for $(L^2_{\mathbb{C}}(\mathbb{R}^3, \varphi_a))^2$ and $\langle\,|\,\rangle_2$ denotes the scalar product for $L^2_{\mathbb{C}}(\mathbb{R}^3, \varphi_a)$. In addition, we note that for every $\lambda \in \mathbb{C}\backslash\mathbb{R}$ by

$$B_{W\lambda} f := \frac{1}{|\ |^2 - \lambda^2}\begin{pmatrix} v_3 + \lambda & v_1 - i\,v_2 \\ v_1 + i\,v_2 & -(v_3 - \lambda) \end{pmatrix} \cdot \begin{pmatrix} f_1 \\ f_2 \end{pmatrix} ,$$

for every $f \in (L^2_{\mathbb{C}}(\mathbb{R}^3, \varphi_a))^2$, there is defined a bounded linear operator on $(L^2_{\mathbb{C}}(\mathbb{R}^3, \varphi_a))^2$ such that $B_{W\lambda} f \in (D(T_{v_1}) \cap D(T_{v_2}) \cap D(T_{v_3}))^2$ and

$$\begin{aligned}
&(A_W - \lambda) B_{W\lambda} f \\
&= \begin{pmatrix} v_3 - \lambda & v_1 - i\,v_2 \\ v_1 + i\,v_2 & -(v_3 + \lambda) \end{pmatrix} \cdot \frac{1}{|\ |^2 - \lambda^2}\begin{pmatrix} v_3 + \lambda & v_1 - i\,v_2 \\ v_1 + i\,v_2 & -(v_3 - \lambda) \end{pmatrix} \cdot \begin{pmatrix} f_1 \\ f_2 \end{pmatrix} \\
&= \frac{1}{|\ |^2 - \lambda^2}\begin{pmatrix} |\ |^2 - \lambda^2 & 0 \\ 0 & |\ |^2 - \lambda^2 \end{pmatrix} \cdot \begin{pmatrix} f_1 \\ f_2 \end{pmatrix} = \begin{pmatrix} f_1 \\ f_2 \end{pmatrix} ,
\end{aligned}$$

for every $f \in (L^2_{\mathbb{C}}(\mathbb{R}^3, \varphi_a))^2$. Hence, see, e.g., Corollary 12.4.10 in the Appendix of [7], A_W is self-adjoint, and it follows that

$$(A_W - \lambda)^{-1} = B_{W\lambda} \ .$$

As a consequence, by

$$A_W^2 f := A_W A_W f \ ,$$

for every $f \in D(A_W^2) = \{f \in D(A_W) : A_W f \in D(A_W)\}$, where $D(A_W)$ denotes the domain of A_W, there is defined a densely, linear and positive self-adjoint operator A_W^2 in $(L^2_{\mathbb{C}}(\mathbb{R}^3, \varphi_a))^2$. In particular, we have that

$$A_W^2 f = \begin{pmatrix} v_3 & v_1 - i\,v_2 \\ v_1 + i\,v_2 & -v_3 \end{pmatrix} \cdot \begin{pmatrix} v_3 & v_1 - i\,v_2 \\ v_1 + i\,v_2 & -v_3 \end{pmatrix} \cdot \begin{pmatrix} f_1 \\ f_2 \end{pmatrix}$$
$$= \begin{pmatrix} |\ |^2 f_1 \\ |\ |^2 f_2 \end{pmatrix} = \begin{pmatrix} T_{|\ |^2} f_1 \\ T_{|\ |^2} f_2 \end{pmatrix} , \tag{1.26}$$

for every $f \in D(A_W^2)$, where $T_{|\ |^2}$ denotes the maximal multiplication operator in $L^2_{\mathbb{C}}(\mathbb{R}^3, \varphi_a)$ with $|\ |^2$. We note that if $f, g \in D(A_W^2)$, then the paths $u, v : \mathbb{R} \to (L^2_{\mathbb{C}}(\mathbb{R}^3, \varphi_a))^2$, for every $t \in \mathbb{R}$ defined by

$$u(t) := e^{itA_W} f \ , \ \ v(t) := e^{-itA_W} g \ ,$$

assume their values in $D(A_W^2)$ and are twice differentiable, with derivatives

$$u''(t) = -A_W^2 u(t) \ , \ \ v''(t) = -A_W^2 v(t) \ , \tag{1.27}$$

for every $t \in \mathbb{R}$, i.e., essentially, the components of u and v satisfy a wave equation, describing the propagation of a massless field in Minkowski space. In this connection, we remind that $L^2_{\mathbb{C}}(\mathbb{R}^3, \varphi_a)$ might be interpreted as momentum space. The observation (1.27) can be used to formulate a well-posed initial value problem for the wave equation

$$u''(t) = -A_W u(t) \ ,$$

$t \in \mathbb{R}$, for general densely-defined, linear and semi-bounded self-adjoint operators A_W in Hilbert spaces. For details, we refer to [4], Theorem 2.2.1 and Corollary 2.2.2.

Since A_W leaves $(C_0(\mathbb{R}^3, \mathbb{C}))^2$ invariant, we have that $(C_0(\mathbb{R}^3, \mathbb{C}))^2$ is part of the C^∞-vectors of A_W,

$$(C_0(\mathbb{R}^3, \mathbb{C}))^2 \subset C^\infty(A_W) \ .$$

In addition, for $f \in (C_0(\mathbb{R}^3, \mathbb{C}))^2$, we have that

$$\begin{aligned} \|A_W f\|^2 &= \|v_3 f_1 + (v_1 - i\,v_2) f_2\|_2^2 + \|(v_1 - i\,v_2) f_1 - v_3 f_2\|_2^2 \\ &\leqslant (\|v_3 f_1\|_2 + \|(v_1 - i\,v_2) f_2\|_2)^2 + (\|(v_1 + i\,v_2) f_1\|_2 + \|v_3 f_2\|_2)^2 \\ &\leqslant 2R^2(\|f_1\|_2 + \|f_2\|_2)^2 \leqslant 4R^2(\|f_1\|_2^2 + \|f_2\|_2^2) = 4R^2\|f\|^2 \end{aligned}$$

and hence that

$$\|A_W f\| \leqslant 2R\,\|f\| \ ,$$

where $R > 0$ is such that $\mathrm{supp}(f_1), \mathrm{supp}(f_2) \subset U_R(0)$. Since A_W leaves the support of the components of f unchanged, from the latter it follows that

$$\|A_W^k f\| \leqslant (2R)^k\,\|f\| \ ,$$

for every $k \in \mathbb{N}$. Hence, for every $t > 0$, the sequence

$$\left(\frac{t^k}{k!}\,A_W^k f\right)_{k\in\mathbb{N}}$$

is absolutely summable, and therefore f is an analytic vector for A_W. As a consequence, for every $t \in \mathbb{R}$, we have that

$$\begin{aligned}
\exp(itA_W)f &= \sum_{k=0}^{\infty} \frac{(it)^k}{k!}\,A_W^k f = \sum_{k=0}^{\infty} \frac{(it)^{2k}}{(2k)!}\,A_W^{2k} f + \sum_{k=0}^{\infty} \frac{(it)^{2k+1}}{(2k+1)!}\,A_W^{2k+1} f \\
&= \sum_{k=0}^{\infty} (-1)^k \frac{t^{2k}}{(2k)!}\,A_W^{2k} f + i \sum_{k=0}^{\infty} (-1)^k \frac{t^{2k+1}}{(2k+1)!}\,A_W^{2k} A_W f \\
&= \sum_{k=0}^{\infty} (-1)^k \frac{t^{2k}}{(2k)!} \begin{pmatrix} |\ |^{2k} f_1 \\ |\ |^{2k} f_2 \end{pmatrix} + i \sum_{k=0}^{\infty} (-1)^k \frac{t^{2k+1}}{(2k+1)!} \begin{pmatrix} |\ |^{2k} (A_W f)_1 \\ |\ |^{2k} (A_W f)_2 \end{pmatrix} \\
&= \begin{pmatrix} \cos(t\,|\ |) f_1 \\ \cos(t\,|\ |) f_2 \end{pmatrix} + i \begin{pmatrix} \frac{\sin(t\,|\ |)}{|\ |}\,(A_W f)_1 \\ \frac{\sin(t\,|\ |)}{|\ |}\,(A_W f)_2 \end{pmatrix} ,
\end{aligned}$$

where we used (1.26). We note that this implies that $\exp(itA_W)f \in (C_0(\mathbb{R}^3, \mathbb{C}))^2$ and therefore that $(C_0(\mathbb{R}^3, \mathbb{C}))^2$ is a core for A_W. Hence, we arrive at the following result.

For every $t \in \mathbb{R}$ and $f \in D(A_W)$, we have that

$$\exp(itA_W)f = \begin{pmatrix} \cos(t\,|\ |) \cdot f_1 \\ \cos(t\,|\ |) \cdot f_2 \end{pmatrix} + i \,.\begin{pmatrix} \frac{\sin(t\,|\ |)}{|\ |} \cdot (A_W f)_1 \\ \frac{\sin(t\,|\ |)}{|\ |} \cdot (A_W f)_2 \end{pmatrix} .$$

In the following, we are going to show that the spectrum $\sigma(A_W)$ of A_W is given by all real numbers. For this purpose, for $\nu \in \mathbb{N}^*$, we define the auxiliary function $k_\nu \in C_0(\mathbb{R}^3, \mathbb{C})$ by

$$k_\nu(v) := \begin{cases} 0 & \text{if } |v| \geqslant 1/\nu, \\ 1 - \nu^2 |v|^2 & \text{if } |v| < 1/\nu. \end{cases}$$

for every $v \in \mathbb{R}^3$. Then $k_\nu(0) = 1$, $0 \leqslant k_\nu \leqslant 1$, $\mathrm{supp}(k_\nu) \subset B_{1/\nu}(0)$. Hence for $\mathfrak{v}_0 \in \mathbb{R}^3 \backslash \{0\}$, we have that

$$f_{\mathfrak{v}_0,\nu} := k_\nu(\mathrm{id}_{\mathbb{R}^3} - \mathfrak{v}_0) \in C_0(\mathbb{R}^3, \mathbb{C}) ,$$

$f_{\mathfrak{v}_0,\nu}(\mathfrak{v}_0) = 1$, $0 \leqslant f_{\mathfrak{v}_0,\nu} \leqslant 1$, $\mathrm{supp}(f_{\mathfrak{v}_0,\nu}) \subset B_{1/\nu}(\mathfrak{v}_0)$. In the following, let $\lambda \in \mathbb{R}\backslash\{0\}$ and $\mathfrak{v}_0 \in \mathbb{R}^3\backslash\{0\}$ such that

$$\begin{cases} |\mathfrak{v}_0| = \lambda & \text{if } \lambda > 0 \\ |\mathfrak{v}_0| = -\lambda & \text{if } \lambda < 0 \end{cases} .$$

If $\lambda > 0$, we have that

$$
\begin{aligned}
&(A_W-\lambda)\left[f_{\mathfrak{v}_0,\nu}\cdot\begin{pmatrix} v_3+|\ | \\ v_1+i\,v_2\end{pmatrix}\right]=f_{\mathfrak{v}_0,\nu}\cdot\begin{pmatrix} v_3-\lambda & v_1-i\,v_2 \\ v_1+i\,v_2 & -(v_3+\lambda)\end{pmatrix}\cdot\begin{pmatrix} v_3+|\ | \\ v_1+i\,v_2\end{pmatrix}\\
&=f_{\mathfrak{v}_0,\nu}\cdot\begin{pmatrix} |\ |^2-\lambda v_3+v_3|\ |-\lambda|\ | \\ (|\ |-\lambda)\,(v_1+i\,v_2)\end{pmatrix}=-f_{\mathfrak{v}_0,\nu}\cdot(\lambda-|\ |)\cdot\begin{pmatrix} v_3+|\ | \\ v_1+i\,v_2\end{pmatrix}\\
&=f_{\mathfrak{v}_0,\nu}\cdot(|\ |-|\mathfrak{v}_0|)\cdot\begin{pmatrix} v_3+|\ | \\ v_1+i\,v_2\end{pmatrix}\,,
\end{aligned}
$$

whereas if $\lambda<0$, we have that

$$
\begin{aligned}
&(A_W-\lambda)\left[f_{\mathfrak{v}_0,\nu}\cdot\begin{pmatrix} v_3-|\ | \\ v_1+i\,v_2\end{pmatrix}\right]=f_{\mathfrak{v}_0,\nu}\cdot\begin{pmatrix} v_3-\lambda & v_1-i\,v_2 \\ v_1+i\,v_2 & -(v_3+\lambda)\end{pmatrix}\cdot\begin{pmatrix} v_3-|\ | \\ v_1+i\,v_2\end{pmatrix}\\
&=f_{\mathfrak{v}_0,\nu}\cdot\begin{pmatrix} |\ |^2-\lambda v_3-v_3|\ |+\lambda|\ | \\ -(|\ |+\lambda)\,(v_1+i\,v_2)\end{pmatrix}=f_{\mathfrak{v}_0,\nu}\cdot\begin{pmatrix} (|\ |+\lambda)(|\ |-v_3) \\ -(|\ |+\lambda)\,(v_1+i\,v_2)\end{pmatrix}\\
&=-f_{\mathfrak{v}_0,\nu}\cdot(\lambda+|\ |)\cdot\begin{pmatrix} v_3-|\ | \\ v_1+i\,v_2\end{pmatrix}=-f_{\mathfrak{v}_0,\nu}\cdot(|\ |-|\mathfrak{v}_0|)\cdot\begin{pmatrix} v_3-|\ | \\ v_1+i\,v_2\end{pmatrix}\,.
\end{aligned}
$$

Since for $v\in B_{1/\nu}(\mathfrak{v}_0)$, we have that

$$
\big|\,|v|-|\mathfrak{v}_0|\,\big|\leqslant|v-\mathfrak{v}_0|\leqslant\frac{1}{\nu}\,,
$$

we infer that

$$
\begin{aligned}
&\left\|f_{\mathfrak{v}_0,\nu}\cdot(|\ |-|\mathfrak{v}_0|)\begin{pmatrix} v_3\pm|\ | \\ v_1+i\,v_2\end{pmatrix}\right\|^2\\
&=\int_{\mathbb{R}^3}(v_a^0)^{-1}\big|\,|\ |-|\mathfrak{v}_0|\,\big|^2\left[(v_3\pm|\ |)^2+v_1^2+v_2^2\right]|f_{\mathfrak{v}_0,\nu}|^2dv^3\\
&\leqslant\frac{1}{\nu^2}\int_{B_{1/\nu}(\mathfrak{v}_0)}(v_a^0)^{-1}\left[(v_3\pm|\ |)^2+v_1^2+v_2^2\right]|f_{\mathfrak{v}_0,\nu}|^2dv^3\\
&=\frac{1}{\nu^2}\left\|f_{\mathfrak{v}_0,\nu}\cdot\begin{pmatrix} v_3+|\ | \\ v_1+i\,v_2\end{pmatrix}\right\|^2
\end{aligned}
$$

and hence that

$$
\lim_{\nu\to\infty}\frac{\left\|(A_W-\lambda)\left[f_{\mathfrak{v}_0,\nu}\cdot\begin{pmatrix} v_3\pm|\ | \\ v_1+i\,v_2\end{pmatrix}\right]\right\|}{\left\|f_{\mathfrak{v}_0,\nu}\cdot\begin{pmatrix} v_3\pm|\ | \\ v_1+i\,v_2\end{pmatrix}\right\|}=0\,.
$$

Therefore, see, e.g., Theorem 12.5.3 in the Appendix of [7], it follows that $\lambda\in\sigma(A_W)$. As a consequence, $\mathbb{R}\backslash\{0\}\subset\sigma(A_W)$ and, since $\sigma(A_W)$ is a closed subset of $\mathbb{R}$, that $\sigma(A_W)=\mathbb{R}$.

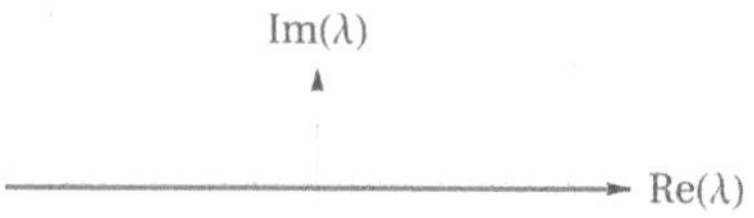

Fig. 1.7 Depiction of the spectrum (in red) of H_R. The spectra of H_R and H_L coincide

> The spectrum of A_W is given by $\sigma(A_W) = \mathbb{R}$.

Therefore, the spectra of the Hamilton operators H_L and H_R are given by all real numbers and hence not bounded from below. As a consequence, like the Dirac equation later, Weyl's equations are not suitable for the definition of a one-particle theory (Fig. 1.7).

1.8 Dirac Spinors

If $\hbar_1, \hbar_2 : \mathrm{SL}(2, \mathbb{C}) \to \mathrm{SL}(2, \mathbb{C})$ are homomorphisms, then we define the direct sum $D^{\hbar_1} \oplus D^{\hbar_2}$ of $D^{\hbar_1}$ and $D^{\hbar_2}$ by

$$(D^{\hbar_1} \oplus D^{\hbar_2})(G) :=$$
$$\left(\mathbb{C}^4 \to \mathbb{C}^4, \psi \mapsto \begin{pmatrix} [\hbar_1(G)]_{11} & [\hbar_1(G)]_{12} & 0 & 0 \\ [\hbar_1(G)]_{21} & [\hbar_1(G)]_{22} & 0 & 0 \\ 0 & 0 & [\hbar_2(G)]_{11} & [\hbar_2(G)]_{12} \\ 0 & 0 & [\hbar_2(G)]_{21} & [\hbar_2(G)]_{22} \end{pmatrix} \cdot \psi \right),$$

for every $G \in \mathrm{SL}(2, \mathbb{C})$. From the linearity of matrix multiplication, it follows that $D^{\hbar}(G)$ is linear, for every $G \in \mathrm{SL}(2, \mathbb{C})$.

> Further,
>
> $$D^{\hbar_1} \oplus D^{\hbar_2} := \begin{pmatrix} \mathrm{SL}(2, \mathbb{C}) \to & L(\mathbb{C}^4) \\ G & \mapsto (D^{\hbar_1} \oplus D^{\hbar_2})(G) \end{pmatrix},$$
>
> is a representation of $\mathrm{SL}(2, \mathbb{C})$,

since for $G_1, G_2 \in \mathrm{SL}(2, \mathbb{C})$, we have that

$$
\begin{aligned}
&(D^{\hbar_1} \oplus D^{\hbar_2})(G_1 G_2)\psi \\
&= \begin{pmatrix} [\hbar_1(G_1G_2)]_{11} & [\hbar_1(G_1G_2)]_{12} & 0 & 0 \\ [\hbar_1(G_1G_2)]_{21} & [\hbar_1(G_1G_2)]_{22} & 0 & 0 \\ 0 & 0 & [\hbar_2(G_1G_2)]_{11} & [\hbar_2(G_1G_2)]_{12} \\ 0 & 0 & [\hbar_2(G_1G_2)]_{21} & [\hbar_2(G_1G_2)]_{22} \end{pmatrix} \cdot \psi \\
&= \begin{pmatrix} [\hbar_1(G_1)\cdot\hbar_1(G_2)]_{11} & [\hbar_1(G_1)\cdot\hbar_1(G_2)]_{12} & 0 & 0 \\ [\hbar_1(G_1)\cdot\hbar_1(G_2)]_{21} & [\hbar_1(G_1)\cdot\hbar_1(G_2)]_{22} & 0 & 0 \\ 0 & 0 & [\hbar_2(G_1)\cdot\hbar_2(G_2)]_{11} & [\hbar_2(G_1)\cdot\hbar_2(G_2)]_{12} \\ 0 & 0 & [\hbar_2(G_1)\cdot\hbar_2(G_2)]_{21} & [\hbar_2(G_1)\cdot\hbar_2(G_2)]_{22} \end{pmatrix} \cdot \psi \\
&= \begin{pmatrix} \sum_{j=1}^{2}[\hbar_1(G_1)]_{1j}\cdot[\hbar_1(G_2)]_{j1} & \sum_{j=1}^{2}[\hbar_1(G_1)]_{1j}\cdot[\hbar_1(G_2)]_{j2} & 0 & 0 \\ \sum_{j=1}^{2}[\hbar_1(G_1)]_{2j}\cdot[\hbar_1(G_2)]_{j1} & \sum_{j=1}^{2}[\hbar_1(G_1)]_{2j}\cdot[\hbar_1(G_2)]_{j2} & 0 & 0 \\ 0 & 0 & \sum_{j=1}^{2}[\hbar_2(G_1)]_{1j}\cdot[\hbar_2(G_2)]_{j1} & \sum_{j=1}^{2}[\hbar_2(G_1)]_{1j}\cdot[\hbar_2(G_2)]_{j2} \\ 0 & 0 & \sum_{j=1}^{2}[\hbar_2(G_1)]_{2j}\cdot[\hbar_2(G_2)]_{j1} & \sum_{j=1}^{2}[\hbar_2(G_1)]_{2j}\cdot[\hbar_2(G_2)]_{j2} \end{pmatrix} \cdot \psi \\
&= \begin{pmatrix} [\hbar_1(G_1)]_{11} & [\hbar_1(G_1)]_{12} & 0 & 0 \\ [\hbar_1(G_1)]_{21} & [\hbar_1(G_1)]_{22} & 0 & 0 \\ 0 & 0 & [\hbar_2(G_1)]_{11} & [\hbar_2(G_1)]_{12} \\ 0 & 0 & [\hbar_2(G_1)]_{21} & [\hbar_2(G_1)]_{22} \end{pmatrix} \cdot \begin{pmatrix} [\hbar_1(G_2)]_{11} & [\hbar_1(G_2)]_{12} & 0 & 0 \\ [\hbar_1(G_2)]_{21} & [\hbar_1(G_2)]_{22} & 0 & 0 \\ 0 & 0 & [\hbar_2(G_2)]_{11} & [\hbar_2(G_2)]_{12} \\ 0 & 0 & [\hbar_2(G_2)]_{21} & [\hbar_2(G_2)]_{22} \end{pmatrix} \cdot \psi \\
&= (D^{\hbar_1} \oplus D^{\hbar_2})(G_1)((D^{\hbar_1} \oplus D^{\hbar_2})(G_2)\,\psi) = ((D^{\hbar_1} \oplus D^{\hbar_2})(G_1) \circ (D^{\hbar_1} \oplus D^{\hbar_2})(G_2))\,\psi\ ,
\end{aligned}
$$

$$
(D^{\hbar_1} \oplus D^{\hbar_2})(E_{2\times 2})\psi = \begin{pmatrix} [\hbar_1(E_{2\times 2})]_{11} & [\hbar_1(E_{2\times 2})]_{12} & 0 & 0 \\ [\hbar_1(E_{2\times 2})]_{21} & [\hbar_1(E_{2\times 2})]_{22} & 0 & 0 \\ 0 & 0 & [\hbar_2(E)]_{11} & [\hbar_2(E_{2\times 2})]_{12} \\ 0 & 0 & [\hbar_2(E_{2\times 2})]_{21} & [\hbar_2(E_{2\times 2})]_{22} \end{pmatrix} \cdot \psi = E_{4\times 4} \cdot \psi = \psi\ ,
$$

for every $\psi \in \mathbb{C}^4$ and hence that

$$
\begin{aligned}
&(D^{\hbar_1} \oplus D^{\hbar_2})(G_1 G_2) = (D^{\hbar_1} \oplus D^{\hbar_2})(G_1) \circ (D^{\hbar_1} \oplus D^{\hbar_2})(G_2)\ , \\
&(D^{\hbar_1} \oplus D^{\hbar_2})(E) = \mathrm{id}_{\mathbb{C}^4}\ .
\end{aligned}
$$

We note that $D^{\hbar_1} \oplus D^{\hbar_2}$ leaves the subspaces

$$
\{{}^t(\psi_1, \psi_2, 0, 0) : \psi_1, \psi_2 \in \mathbb{C}\}\ ,\ \ \{{}^t(0, 0, \psi_3, \psi_4) : \psi_3, \psi_4 \in \mathbb{C}\}
$$

of $\mathbb{C}^4$ invariant and hence is reducible.

The elements of the representation space $\mathbb{C}^4$ of $D^{\hbar_L} \oplus D^{\hbar_R}$ are $\mathrm{SL}(2, \mathbb{C})$ four-component ("Dirac"-) spinors. We note that, from the representations (1.14) and (1.15), for $G \in \mathrm{SL}(2, \mathbb{C})$, it follows that

$$
\|(D^{\hbar_L} \oplus D^{\hbar_R})(G)\|_\infty = \|G\|_\infty\ ,
$$

where we define

$$
\|G\|_\infty := \max\{|G_{11}|, |G_{12}|, |G_{21}|, |G_{22}|\}\ ,
$$

for every $G \in \mathrm{M}(2, \mathbb{C})$. In addition, from the same argument, it follows that $D^{\hbar_L} \oplus D^{\hbar_R}$ is continuous, i.e., if $G_1, G_2, \ldots$ is a sequence in $\mathrm{SL}(2, \mathbb{C})$ that converges component-wise to $G \in \mathrm{SL}(2,\mathbb{C})$, then the sequence $(D^{\hbar_L} \oplus D^{\hbar_R})(G_1), (D^{\hbar_L} \oplus D^{\hbar_R})(G_2), \ldots$ converges component-wise to $(D^{\hbar_L} \oplus D^{\hbar_R})(G)$. We note that, in particular, we have for every

$$\begin{pmatrix} \alpha & \beta \\ \gamma & \delta \end{pmatrix} \in \mathrm{SL}(2,\mathbb{C})$$

that

$$(D^{\hbar_L} \oplus D^{\hbar_R})\Big(\begin{pmatrix} \alpha & \beta \\ \gamma & \delta \end{pmatrix}\Big) =$$

$$\left(\mathbb{C}^4 \to \mathbb{C}^4, \psi \mapsto \begin{pmatrix} \delta^* & -\gamma^* & 0 & 0 \\ -\beta^* & \alpha^* & 0 & 0 \\ 0 & 0 & \alpha & \beta \\ 0 & 0 & \gamma & \delta \end{pmatrix} \cdot \psi \right),$$

for every $\psi \in \mathbb{C}^4$.

1.9 Dirac Representation of SL(2, ℂ)

In the following, let $a \geqslant 0$. For $G \in \mathrm{SL}(2,\mathbb{C})$, we define $(T_a^{\hbar_L} \oplus T_a^{\hbar_R})(G) : (L^2_{\mathbb{C}}(\mathbb{R}^3, \varphi_a))^4 \to (L^2_{\mathbb{C}}(\mathbb{R}^3, \varphi_a))^4$ by

$$\begin{aligned}(T_a^{\hbar_L} \oplus T_a^{\hbar_R})(G) f := {}^t\big(&[T_a^{\hbar_L}(G)\,{}^t(f_1, f_2)]_1, [T_a^{\hbar_L}(G)\,{}^t(f_1, f_2)]_2, \\ &[T_a^{\hbar_R}(G)\,{}^t(f_3, f_4)]_1, [T_a^{\hbar_R}(G)\,{}^t(f_3, f_4)]_2\big)\,,\end{aligned}$$

for every $f \in (L^2_{\mathbb{C}}(\mathbb{R}^3, \varphi_a))^4$. From the linearity of $T_a^{\hbar_L}(G)$ and $T_a^{\hbar_R}(G)$ and the linearity of the coordinate projections of $(L^2_{\mathbb{C}}(\mathbb{R}^3, \varphi_a))^2$, it follows the linearity of $(T_a^{\hbar_L} \oplus T_a^{\hbar_R})(G)$. Further, since

$$\begin{aligned}\|(T_a^{\hbar_L} \oplus T_a^{\hbar_R})(G) f\|^2 &= \|[T_a^{\hbar_L}(G)\,{}^t(f_1, f_2)]_1\|_2^2 + \|[T_a^{\hbar_L}(G)\,{}^t(f_1, f_2)]_2\|_2^2 \\ &+ \|[T_a^{\hbar_R}(G)\,{}^t(f_3, f_4)]_1\|_2^2 + \|[T_a^{\hbar_R}(G)\,{}^t(f_3, f_4)]_2\|_2^2 \\ &\leqslant 4\,\|G\|_\infty^2(\|f_1\|_2^2 + \|f_2\|_2^2) + 4\,\|G\|_\infty^2(\|f_3\|_2^2 + \|f_4\|_2^2) = 4\,\|G\|_\infty^2\|f\|^2\,,\end{aligned}$$

for every $f \in (L^2_{\mathbb{C}}(\mathbb{R}^3, \varphi_a))^4$, where $\| \ \|$ denotes the norm that is induced by the scalar product $\langle \,|\, \rangle$ on $(L^2_{\mathbb{C}}(\mathbb{R}^3, \varphi_a))^4$ and $\| \ \|_2$ denotes the norm that is induced by the scalar product $\langle \,|\, \rangle_2$ on $L^2_{\mathbb{C}}(\mathbb{R}^3, \varphi_a)$. It follows that $(T_a^{\hbar_L} \oplus T_a^{\hbar_R})(G) \in L(((L^2_{\mathbb{C}}(\mathbb{R}^3, \varphi_a))^4)$. In particular, if $G \in \mathrm{SU}(2)$, then

$$
\begin{aligned}
&\langle (T_a^{\hbar_L} \oplus T_a^{\hbar_R})(G) f | (T_a^{\hbar_L} \oplus T_a^{\hbar_R})(G) g \rangle \\
&= \langle [T_a^{\hbar_L}(G)\,{}^t(f_1, f_2)]_1 | [T_a^{\hbar_L}(G)\,{}^t(g_1, g_2)]_1 \rangle_2 \\
&\quad + \langle [T_a^{\hbar_L}(G)\,{}^t(f_1, f_2)]_2 | [T_a^{\hbar_L}(G)\,{}^t(g_1, g_2)]_2 \rangle_2 \\
&\quad + \langle [T_a^{\hbar_R}(G)\,{}^t(f_3, f_4)]_1 | [T_a^{\hbar_R}(G)\,{}^t(g_3, g_4)]_1 \rangle_2 \\
&\quad + \langle [T_a^{\hbar_R}(G)\,{}^t(f_3, f_4)]_2 | [T_a^{\hbar_R}(G)\,{}^t(g_3, g_4)]_2 \rangle_2 \\
&= \langle f_1 | g_1 \rangle_2 + \langle f_2 | g_2 \rangle_2 + \langle f_3 | g_3 \rangle_2 + \langle f_4 | g_4 \rangle_2 = \langle f | g \rangle \ ,
\end{aligned}
$$

for all $f, g \in (L^2_{\mathbb{C}}(\mathbb{R}^3, \varphi_a))^4$, and hence $(T_a^{\hbar_L} \oplus T_a^{\hbar_R})(G)$ preserves the scalar product. As a consequence, by the direct sum $T_a^{\hbar_L} \oplus T_a^{\hbar_R}$ of $T_a^{\hbar_L}$ and $T_a^{\hbar_R}$ that associates with every $G \in \mathrm{SL}(2, \mathbb{C})$ the corresponding map $(T_a^{\hbar_L} \oplus T_a^{\hbar_R})(G)$, there is defined a map

$$
T_a^{\hbar_L} \oplus T_a^{\hbar_R} : \mathrm{SL}(2, \mathbb{C}) \to L((L^2_{\mathbb{C}}(\mathbb{R}^3, \varphi_a))^4, (L^2_{\mathbb{C}}(\mathbb{R}^3, \varphi_a))^4) \ .
$$

Further, for $G, H \in \mathrm{SL}(2, \mathbb{C})$ and $f \in (L^2_{\mathbb{C}}(\mathbb{R}^3, \varphi_a))^4$, we have that

$$
\begin{aligned}
&(T_a^{\hbar_L} \oplus T_a^{\hbar_R})(G)\,(T_a^{\hbar_L} \oplus T_a^{\hbar_R})(H) f \\
&= (T_a^{\hbar_L} \oplus T_a^{\hbar_R})(G)\,{}^t([T_a^{\hbar_L}(H)\,{}^t(f_1, f_2)]_1, [T_a^{\hbar_L}(H)\,{}^t(f_1, f_2)]_2, \\
&\qquad\qquad [T_a^{\hbar_R}(H)\,{}^t(f_3, f_4)]_1, [T_a^{\hbar_R}(H)\,{}^t(f_3, f_4)]_2) \\
&= {}^t([T_a^{\hbar_L}(G)\,{}^t([T_a^{\hbar_L}(H)\,{}^t(f_1, f_2)]_1, [T_a^{\hbar_L}(H)\,{}^t(f_1, f_2)]_2)]_1, \\
&\quad [T_a^{\hbar_L}(G)\,{}^t([T_a^{\hbar_L}(H)\,{}^t(f_1, f_2)]_1, [T_a^{\hbar_L}(H)\,{}^t(f_1, f_2)]_2)]_2, \\
&\quad [T_a^{\hbar_R}(G)\,{}^t([T_a^{\hbar_R}(H)\,{}^t(f_3, f_4)]_1, [T_a^{\hbar_R}(H)\,{}^t(f_3, f_4)]_2)]_1, \\
&\quad [T_a^{\hbar_R}(G)\,{}^t([T_a^{\hbar_R}(H)\,{}^t(f_3, f_4)]_1, [T_a^{\hbar_R}(H)\,{}^t(f_3, f_4)]_2)]_2) \\
&= {}^t([T_a^{\hbar_L}(G) T_a^{\hbar_L}(H)\,{}^t(f_1, f_2)]_1, [T_a^{\hbar_L}(G) T_a^{\hbar_L}(H)\,{}^t(f_1, f_2)]_2, \\
&\quad [T_a^{\hbar_R}(G) T_a^{\hbar_R}(H)\,{}^t(f_3, f_4)]_1, [T_a^{\hbar_R}(G) T_a^{\hbar_R}(H)\,{}^t(f_3, f_4)]_2) \\
&= {}^t([T_a^{\hbar_L}(G \cdot H)\,{}^t(f_1, f_2)]_1, [T_a^{\hbar_L}(G \cdot H)\,{}^t(f_1, f_2)]_2, \\
&\quad [T_a^{\hbar_R}(G \cdot H)\,{}^t(f_3, f_4)]_1, [T_a^{\hbar_R}(G \cdot H)\,{}^t(f_3, f_4)]_2) \\
&= (T_a^{\hbar_L} \oplus T_a^{\hbar_R})(G \cdot H) f
\end{aligned}
$$

and hence that

$$
(T_a^{\hbar_L} \oplus T_a^{\hbar_R})(G \cdot H) = (T_a^{\hbar_L} \oplus T_a^{\hbar_R})(G) \circ (T_a^{\hbar_L} \oplus T_a^{\hbar_R})(H) \ .
$$

Since

$$
\begin{aligned}
&(T_a^{\hbar_L} \oplus T_a^{\hbar_R})(E_{2\times 2}) f \\
&= {}^t([T_a^{\hbar_L}(E_{2\times 2})\,{}^t(f_1, f_2)]_1, [T_a^{\hbar_L}(E_{2\times 2})\,{}^t(f_1, f_2)]_2, \\
&\quad [T_a^{\hbar_R}(E_{2\times 2})\,{}^t(f_3, f_4)]_1, [T_a^{\hbar_R}(E_{2\times 2})\,{}^t(f_3, f_4)]_2) = f \ ,
\end{aligned}
$$

for every $f \in (L^2_{\mathbb{C}}(\mathbb{R}^3, \varphi_a))^4$, $(T_a^{\hbar_L} \oplus T_a^{\hbar_R})(E)$ is given by the identity map on $(L^2_{\mathbb{C}}(\mathbb{R}^3, \varphi_a))^4$. Hence, it follows that $T_a^{\hbar_L} \oplus T_a^{\hbar_R}$ is a representation of $\mathrm{SL}(2, \mathbb{C})$. We note that this implies also that $(T_a^{\hbar_L} \oplus T_a^{\hbar_R})(G)$ is unitary for every $G \in \mathrm{SU}(2)$. Further, if $G, H \in \mathrm{SL}(2, \mathbb{C})$ and $f \in (L^2_{\mathbb{C}}(\mathbb{R}^3, \varphi_a))^4$, we have that

$$
\begin{aligned}
&(T_a^{\hbar_L} \oplus T_a^{\hbar_R})(G)f - (T_a^{\hbar_L} \oplus T_a^{\hbar_R})(H)f \\
&= {}^t([(T_a^{\hbar_L}(G) - T_a^{\hbar_L}(H))\,{}^t(f_1, f_2)]_1, [(T_a^{\hbar_L}(G) - T_a^{\hbar_L}(H))\,{}^t(f_1, f_2)]_2, \\
&\quad [(T_a^{\hbar_R}(G) - T_a^{\hbar_R}(H))\,{}^t(f_3, f_4)]_1, [(T_a^{\hbar_R}(G) - T_a^{\hbar_R}(H))\,{}^t(f_3, f_4)]_2)
\end{aligned}
$$

Since $T_a^{\hbar_L}$ and $T_a^{\hbar_R}$ are strongly continuous, it follows that $T_a^{\hbar_L} \oplus T_a^{\hbar_R}$ is strongly continuous, i.e., if $G_1, G_2, \ldots$ is a sequence in $\mathrm{SL}(2, \mathbb{C})$ that converges component-wise to $G \in \mathrm{SL}(2, \mathbb{C})$, then

$$
\lim_{\nu \to \infty} \|[(T_a^{\hbar_L} \oplus T_a^{\hbar_R})(G_\nu) - (T_a^{\hbar_L} \oplus T_a^{\hbar_R})(G)]f\| = 0 \, ,
$$

for every $f \in (L^2_{\mathbb{C}}(\mathbb{R}^3, \varphi_a))^4$.

Hence, we arrive at the following result. If $a \geqslant 0$, then

$$
T_a^{\hbar_L} \oplus T_a^{\hbar_R} : \mathrm{SL}(2, \mathbb{C}) \to L((L^2_{\mathbb{C}}(\mathbb{R}^3, \varphi_a))^4, (L^2_{\mathbb{C}}(\mathbb{R}^3, \varphi_a))^4) \, ,
$$

defined by

$$
\begin{aligned}
(T_a^{\hbar_L} \oplus T_a^{\hbar_R})(G)f := {}^t(&[T_a^{\hbar_L}(G)\,{}^t(f_1, f_2)]_1, [T_a^{\hbar_L}(G)\,{}^t(f_1, f_2)]_2, \\
&[T_a^{\hbar_R}(G)\,{}^t(f_3, f_4)]_1, [T_a^{\hbar_R}(G)\,{}^t(f_3, f_4)]_2) \, ,
\end{aligned}
$$

for every $G \in \mathrm{SL}(2, \mathbb{C})$ and $f \in (L^2_{\mathbb{C}}(\mathbb{R}^3, \varphi_a))^4$, is a representation of $\mathrm{SL}(2, \mathbb{C})$. In addition, $T_a^{\hbar_L} \oplus T_a^{\hbar_R}$ is strongly continuous, i.e., if $G_1, G_2, \ldots$ is a sequence in $\mathrm{SL}(2, \mathbb{C})$ that converges component-wise to $G \in \mathrm{SL}(2, \mathbb{C})$, then

$$
\lim_{\nu \to \infty} \|[(T_a^{\hbar_L} \oplus T_a^{\hbar_R})(G_\nu) - (T_a^{\hbar_L} \oplus T_a^{\hbar_R})(G)]f\| = 0 \, ,
$$

for every $f \in (L^2_{\mathbb{C}}(\mathbb{R}^3, \varphi_a))^4$. Further, $(T_a^{\hbar_L} \oplus T_a^{\hbar_R})(G)$ is unitary, for every $G \in \mathrm{SU}(2, \mathbb{C})$.

1.9.1 Generators Associated with Rotations

Further, if

$$G : (\mathbb{R}, +) \to \mathrm{SL}(2, \mathbb{C})$$

is one-parameter group, i.e., such that

$$G(0) = E \ , \ \ G(s_1 + s_2) = G(s_1) \cdot G(s_2) \ ,$$

for all $s_1, s_2 \in \mathbb{R}$ and such that, for every sequence $s_1, s_2, \ldots$ in $\mathbb{R}$ that is convergent to $s \in \mathbb{R}$, the corresponding sequence $G(s_1), G(s_2), \ldots$ converges component-wise to $G(s)$, then $(T_a^{\hbar_L} \oplus T_a^{\hbar_R}) \circ G$ is a strongly continuous one-parameter group. Hence, there is an infinitesimal generator[4] A_G in

$$\boxed{X := (L^2_{\mathbb{C}}(\mathbb{R}^3, \varphi_a))^4} \ ,$$

given by

$$D(A_G) = \{f \in X : \lim_{s \to 0, s \neq 0} \frac{1}{s} \left[((T_a^{\hbar_L} \oplus T_a^{\hbar_R}) \circ G)(s) - \mathrm{id}_X\right] f \text{ exists}\}$$

and for every $f \in D(A_G)$ by[5]

$$A_G f = \frac{1}{i} \lim_{s \to 0, s \neq 0} \frac{1}{s} \left[((T_a^{\hbar_L} \oplus T_a^{\hbar_R}) \circ G)(s) - \mathrm{id}_X\right] f \ .$$

In particular, if $\mathrm{Ran}(G) \subset \mathrm{SU}(2)$, then $(T_a^{\hbar_L} \oplus T_a^{\hbar_R}) \circ G$ is a strongly continuous one-parameter unitary group and, according to Stone's theorem, we have that

$$\exp(isA_G) = ((T_a^{\hbar_L} \oplus T_a^{\hbar_R}) \circ G)(s) \ ,$$

for every $s \in \mathbb{R}$, where A_G is (densely-defined, linear and) self-adjoint. In the following, we define the one-parameter groups $G_1, G_2, G_3 : \mathbb{R} \to \mathrm{SU}(2)$ by

$$G_1(\varphi) := \exp(i\,(\varphi/2)\,\sigma_1) \ , \ \ G_2(\varphi) := \exp(i\,(\varphi/2)\,\sigma_2) \ ,$$
$$G_3(\varphi) := \exp(i\,(\varphi/2)\,\sigma_3) \ ,$$

for every $\varphi \in \mathbb{R}$, where $\sigma_1, \sigma_2, \sigma_3$ are the Pauli spin matrices, see Sect. 1.1. Then, for $j \in \{1, 2, 3\}$, $\varphi \in \mathbb{R}^*$ and $f \in X$, it follows that

---

[4] E.g., see [4], Sect. 4.5.

[5] The following definition is not standard. In a standard definition, the factor $1/i$ in the following definition is replaced by -1, see [4], Sect. 4.5. Hence, in the sense of strongly semigroups of operators, the generator of $(T_a^{\hbar_L} \oplus T_a^{\hbar_R}) \circ G$ is given by $-iA_G$.

$$\frac{1}{i\varphi}\left[((T_a^{\hbar_L} \oplus T_a^{\hbar_R}) \circ G_j)(\varphi) - \mathrm{id}_X\right] f$$
$$= {}^t(\frac{1}{i\varphi}\{[T_a^{\hbar_L}(G_j(\varphi))\,{}^t(f_1, f_2)]_1 - f_1\}, \frac{1}{i\varphi}\{[T_a^{\hbar_L}(G_j(\varphi))\,{}^t(f_1, f_2)]_2 - f_2\},$$
$$\frac{1}{i\varphi}\{[T_a^{\hbar_R}(G_j(\varphi))\,{}^t(f_3, f_4)]_1 - f_3\}, \frac{1}{i\varphi}\{[T_a^{\hbar_R}(G_j(\varphi))\,{}^t(f_3, f_4)]_2 - f_4\}) \ ,$$

and hence that

$$D(A_{G_j}) = (D(\hat{L}_j))^4$$

as well as that

$$A_{G_j} f = {}^t\left(\frac{1}{\hbar}\hat{L}_j f_1 + \frac{1}{2}[\sigma_j \cdot {}^t(f_1, f_2)]_1, \frac{1}{\hbar}\hat{L}_j f_2 + \frac{1}{2}[\sigma_j \cdot {}^t(f_1, f_2)]_2,\right.$$
$$\left.\frac{1}{\hbar}\hat{L}_j f_3 + \frac{1}{2}[\sigma_j \cdot {}^t(f_3, f_4)]_1, \frac{1}{\hbar}\hat{L}_j f_4 + \frac{1}{2}[\sigma_j \cdot {}^t(f_3, f_4)]_2\right) \ ,$$

for every $f \in (D(\hat{L}_j))^4$, where the components of angular momentum $\hat{L}_1, \hat{L}_2, \hat{L}_3$ are defined in Sect. 5.2.1 of Part I, and we use the results of Sect. 1.4.1.

Hence, for $j \in \{1, 2, 3\}$, we define the, densely-defined, linear and self-adjoint, j-th component $\hat{J}_j : (D(\hat{L}_j))^4 \to X$ of total angular momentum by

$$\hat{J}_j f := \hbar A_{G_j} f$$
$$= {}^t\left(\hat{L}_j f_1 + \frac{\hbar}{2}[\sigma_j \cdot {}^t(f_1, f_2)]_1, \hat{L}_j f_2 + \frac{\hbar}{2}[\sigma_j \cdot {}^t(f_1, f_2)]_2,\right.$$
$$\left.\hat{L}_j f_3 + \frac{\hbar}{2}[\sigma_j \cdot {}^t(f_3, f_4)]_1, \hat{L}_j f_4 + \frac{\hbar}{2}[\sigma_j \cdot {}^t(f_3, f_4)]_2\right) \ ,$$

for every $f \in (D(\hat{L}_j))^4$ as well as the, bounded linear and self-adjoint, j-th component $\hat{S}_j : X \to X$ of intrinsic angular momentum by

$$\hat{S}_j f := \frac{\hbar}{2}\,{}^t\left([\sigma_j \cdot {}^t(f_1, f_2)]_1, [\sigma_j \cdot {}^t(f_1, f_2)]_2,\right.$$
$$\left.[\sigma_j \cdot {}^t(f_3, f_4)]_1, [\sigma_j \cdot {}^t(f_3, f_4)]_2\right)$$
$$= \frac{\hbar}{2}\begin{pmatrix} (\sigma_j)_{11} & (\sigma_j)_{12} & 0 & 0 \\ (\sigma_j)_{21} & (\sigma_j)_{22} & 0 & 0 \\ 0 & 0 & (\sigma_j)_{11} & (\sigma_j)_{12} \\ 0 & 0 & (\sigma_j)_{21} & (\sigma_j)_{22} \end{pmatrix} \cdot \begin{pmatrix} f_1 \\ f_2 \\ f_3 \\ f_4 \end{pmatrix} \ ,$$

for every $f \in X$.

Exercise 1.11 Show that the operator $\hat{J}_3$ has a pure point spectrum, consisting of eigenvalues of infinite multiplicity, given by

$$\hbar\left(\mathbb{Z}+\frac{1}{2}\right).$$

We note that the eigenvalues of $\hat{S}_j$ are given by $-\hbar/2$ and $\hbar/2$ as well as that

$$\ker\left(\hat{S}_1+\frac{\hbar}{2}\right)=L^2_{\mathbb{C}}(\mathbb{R}^3,\varphi_a)\cdot\begin{pmatrix}0\\0\\1\\-1\end{pmatrix}+L^2_{\mathbb{C}}(\mathbb{R}^3,\varphi_a)\cdot\begin{pmatrix}1\\-1\\0\\0\end{pmatrix},$$

$$\ker\left(\hat{S}_1-\frac{\hbar}{2}\right)=L^2_{\mathbb{C}}(\mathbb{R}^3,\varphi_a)\cdot\begin{pmatrix}0\\0\\1\\1\end{pmatrix}+L^2_{\mathbb{C}}(\mathbb{R}^3,\varphi_a)\cdot\begin{pmatrix}1\\1\\0\\0\end{pmatrix},$$

$$\ker\left(\hat{S}_2+\frac{\hbar}{2}\right)=L^2_{\mathbb{C}}(\mathbb{R}^3,\varphi_a)\cdot\begin{pmatrix}0\\0\\i\\1\end{pmatrix}+L^2_{\mathbb{C}}(\mathbb{R}^3,\varphi_a)\cdot\begin{pmatrix}i\\1\\0\\0\end{pmatrix},$$

$$\ker\left(\hat{S}_2-\frac{\hbar}{2}\right)=L^2_{\mathbb{C}}(\mathbb{R}^3,\varphi_a)\cdot\begin{pmatrix}0\\0\\i\\-1\end{pmatrix}+L^2_{\mathbb{C}}(\mathbb{R}^3,\varphi_a)\cdot\begin{pmatrix}i\\-1\\0\\0\end{pmatrix},$$

$$\ker\left(\hat{S}_3+\frac{\hbar}{2}\right)=L^2_{\mathbb{C}}(\mathbb{R}^3,\varphi_a)\cdot\begin{pmatrix}0\\0\\0\\1\end{pmatrix}+L^2_{\mathbb{C}}(\mathbb{R}^3,\varphi_a)\cdot\begin{pmatrix}0\\1\\0\\0\end{pmatrix},$$

$$\ker\left(\hat{S}_3-\frac{\hbar}{2}\right)=L^2_{\mathbb{C}}(\mathbb{R}^3,\varphi_a)\cdot\begin{pmatrix}0\\0\\1\\0\end{pmatrix}+L^2_{\mathbb{C}}(\mathbb{R}^3,\varphi_a)\cdot\begin{pmatrix}1\\0\\0\\0\end{pmatrix}.$$

Hence, we obtain the following result (Fig. 1.8).

For $j \in \{1, 2, 3\}$, there is a Hilbert basis of X, consisting of eigenvectors of $\hat{S}_j$, corresponding to the eigenvalues $-\hbar/2$ and $\hbar/2$. Therefore, $\hat{S}_j$ has a pure point spectrum given by

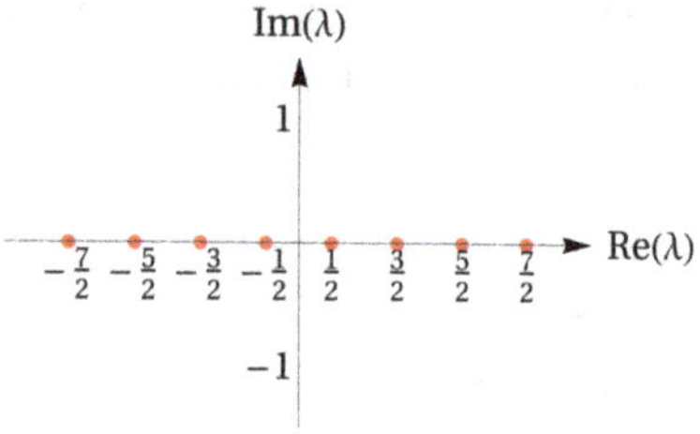

Fig. 1.8 Depiction of the spectrum (in blue) of $\hat{J}_3$

$$\sigma(\hat{S}_j) = \left\{-\frac{\hbar}{2}, \frac{\hbar}{2}\right\} .$$

Further, we note that

$$\hat{S}_j^{2k} = \left(\frac{\hbar}{2}\right)^{2k} .E \ , \ \hat{S}_j^{2k+1} = \left(\frac{\hbar}{2}\right)^{2k} \hat{S}_j \ ,$$

for every $k \in \mathbb{N}$, where we used that

$$\sigma_j^{2k} = E \ , \ \sigma_j^{2k+1} = \sigma_j \ ,$$

for every $k \in \mathbb{N}$. Hence, we have that

$$\begin{aligned}
\exp(\frac{i\varphi}{\hbar}\,\hat{S}_j)f &= \sum_{k=0}^{\infty} \frac{1}{k!}(\frac{i\varphi}{\hbar}\,\hat{S}_j)^k f \\
&= \sum_{k=0}^{\infty} \frac{1}{(2k)!}(\frac{i\varphi}{\hbar}\,\hat{S}_j)^{2k} f + \sum_{k=0}^{\infty} \frac{1}{(2k+1)!}(\frac{i\varphi}{\hbar}\,\hat{S}_j)^{2k+1} f \\
&= \sum_{k=0}^{\infty} \frac{1}{(2k)!}(\frac{i\varphi}{\hbar})^{2k}\,\hat{S}_j^{2k} f + \sum_{k=0}^{\infty} \frac{1}{(2k+1)!}(\frac{i\varphi}{\hbar})^{2k+1}\,\hat{S}_j^{2k+1} f \\
&= \sum_{k=0}^{\infty} \frac{1}{(2k)!}(\frac{i\varphi}{\hbar})^{2k}\left(\frac{\hbar}{2}\right)^{2k} f + \sum_{k=0}^{\infty} \frac{1}{(2k+1)!}(\frac{i\varphi}{\hbar})^{2k+1}\left(\frac{\hbar}{2}\right)^{2k} \hat{S}_j f \\
&= \sum_{k=0}^{\infty} \frac{1}{(2k)!}(\frac{i\varphi}{2})^{2k} f + \sum_{k=0}^{\infty} \frac{1}{(2k+1)!}(\frac{i\varphi}{2})^{2k+1}\,\frac{2}{\hbar}\,\hat{S}_j f \\
&= \cosh(\frac{i\varphi}{2})f + \frac{2}{\hbar}\,\sinh(\frac{i\varphi}{2})\hat{S}_j f = \cos(\frac{\varphi}{2})f + \frac{2i}{\hbar}\,\sin(\frac{\varphi}{2})\,\hat{S}_j f \ ,
\end{aligned}$$

for every $f \in X$ and $\varphi \in \mathbb{R}$. Hence,

$$\boxed{\exp(\frac{i\varphi}{\hbar}\,\hat{S}_j)f = \cos(\frac{\varphi}{2})f + \frac{2i}{\hbar}\,\sin(\frac{\varphi}{2})\,\hat{S}_j f \ ,}$$

for every $f \in X$ and $\varphi \in \mathbb{R}$. As a consequence,

$$\exp\Big(\frac{i\,(\varphi + 2k\pi)}{\hbar}\,\hat{S}_j\Big) f = (-1)^k \exp\Big(\frac{i\varphi}{\hbar}\,\hat{S}_j\Big) f \ ,$$

for every $f \in X$, $\varphi \in \mathbb{R}$ and $k \in \mathbb{Z}$, i.e., an increase of the angle φ about $2k\pi$, $k \in \mathbb{Z}$, results in a multiplication by a phase factor.

1.9.2 Generators Associated with Lorentz Boosts

Again, if

$$G : (\mathbb{R}, +) \rightarrow \mathrm{SL}(2, \mathbb{C})$$

is one-parameter group, i.e., such that

$$G(0) = E \ , \ \ G(s_1 + s_2) = G(s_1) \cdot G(s_2) \ ,$$

for all $s_1, s_2 \in \mathbb{R}$ and such that, for every sequence $s_1, s_2, \dots$ in $\mathbb{R}$ that is convergent to $s \in \mathbb{R}$, the corresponding sequence $G(s_1), G(s_2), \dots$ converges component-wise to $G(s)$, then $(T_a^{\hbar_L} \oplus T_a^{\hbar_R}) \circ G$ is a strongly continuous one-parameter group. Hence, there is an infinitesimal generator[6] A_G in

$$\boxed{X := (L^2_{\mathbb{C}}(\mathbb{R}^3, \varphi_a))^4 \ ,}$$

given by

$$D(A_G) = \{ f \in X : \lim_{s \to 0, s \neq 0} \frac{1}{s} \Big[((T_a^{\hbar_L} \oplus T_a^{\hbar_R}) \circ G)(s) - \mathrm{id}_X \Big] f \text{ exists} \}$$

and for every $f \in D(A_G)$ by

$$A_G f = - \lim_{s \to 0, s \neq 0} \frac{1}{s} \Big[((T_a^{\hbar_L} \oplus T_a^{\hbar_R}) \circ G)(s) - \mathrm{id}_X \Big] f \ .$$

In the following, we define the one-parameter groups $G_1, G_2, G_3 : \mathbb{R} \rightarrow \mathrm{SL}(2, \mathbb{C})$ by

$$G_1(s) := \exp((s/2)\,\sigma_1) \ , \ \ G_2(s) := \exp((s/2)\,\sigma_2) \ ,$$
$$G_3(s) := \exp((s/2)\,\sigma_3) \ ,$$

for every $s \in \mathbb{R}$, where $\sigma_1, \sigma_2, \sigma_3$ are the Pauli spin matrices, see Sect. 1.1. From the continuity of the exponential function, it follows that G_1, G_2 and G_3 are continuous. Then, for $j \in \{1, 2, 3\}$, $s \in \mathbb{R}^*$ and $f \in X$, it follows that

[6] E.g., see [4], Sect. 4.5.

$$
\begin{aligned}
&-\frac{1}{s}\left[((T_a^{\hbar_L}\oplus T_a^{\hbar_R})\circ G_j)(s)-\mathrm{id}_X\right]f\\
&={}^t\Big(-\frac{1}{s}\{[T_a^{\hbar_L}(G_j(s))\,{}^t(f_1,f_2)]_1-f_1\},-\frac{1}{s}\{[T_a^{\hbar_L}(G_j(s))\,{}^t(f_1,f_2)]_2-f_2\},\\
&\qquad -\frac{1}{s}\{[T_a^{\hbar_R}(G_j(s))\,{}^t(f_3,f_4)]_1-f_3\},-\frac{1}{s}\{[T_a^{\hbar_R}(G_j(s))\,{}^t(f_3,f_4)]_2-f_4\}\Big)\ ,
\end{aligned}
$$

and hence that

$$D(A_{G_j})=(D(\hat{L}_{0j}))^4$$

as well as that

$$
\begin{aligned}
A_{G_j}f={}^t\bigg(&-\frac{i}{\hbar}\hat{L}_{0j}f_1+\frac{1}{2}[\sigma_j\cdot{}^t(f_1,f_2)]_1,-\frac{i}{\hbar}\hat{L}_{0j}f_2+\frac{1}{2}[\sigma_j\cdot{}^t(f_1,f_2)]_2,\\
&-\frac{i}{\hbar}\hat{L}_{0j}f_3-\frac{1}{2}[\sigma_j\cdot{}^t(f_3,f_4)]_1,-\frac{i}{\hbar}\hat{L}_{0j}f_4-\frac{1}{2}[\sigma_j\cdot{}^t(f_3,f_4)]_2\bigg)\ ,
\end{aligned}
$$

for every $f\in(D(\hat{L}_{0j}))^4$, where the components of the momenta $\hat{L}_{01},\hat{L}_{02},\hat{L}_{03}$ are defined in Sect. 5.2.2 of Part I, and we use the results of Sect. 1.4.1.2.

Hence, for $j\in\{1,2,3\}$, by

$$
\begin{aligned}
\hat{J}_{0j}f&:=i\hbar A_{G_j}f\\
&={}^t\bigg(\hat{L}_{0j}f_1+\frac{\hbar}{2}[i\sigma_j\cdot{}^t(f_1,f_2)]_1,\hat{L}_jf_2+\frac{\hbar}{2}[i\sigma_j\cdot{}^t(f_1,f_2)]_2,\\
&\qquad \hat{L}_{0j}f_3-\frac{\hbar}{2}[i\sigma_j\cdot{}^t(f_3,f_4)]_1,\hat{L}_{0j}f_4-\frac{\hbar}{2}[i\sigma_j\cdot{}^t(f_3,f_4)]_2\bigg)\\
&=\begin{pmatrix}\hat{L}_{0j}f_1\\ \hat{L}_{0j}f_2\\ \hat{L}_{0j}f_3\\ \hat{L}_{0j}f_4\end{pmatrix}+\frac{i\hbar}{2}\begin{pmatrix}(\sigma_j)_{11}&(\sigma_j)_{12}&0&0\\(\sigma_j)_{21}&(\sigma_j)_{22}&0&0\\0&0&-(\sigma_j)_{11}&-(\sigma_j)_{12}\\0&0&-(\sigma_j)_{21}&-(\sigma_j)_{22}\end{pmatrix}\cdot\begin{pmatrix}f_1\\f_2\\f_3\\f_4\end{pmatrix}\ ,
\end{aligned}
$$

for every $f\in(D(\hat{L}_{0j}))^4$, there is given a densely-defined, linear operator in $(L^2_{\mathbb{C}}(\mathbb{R}^3,\varphi_a))^4$, such that $-(i/\hbar)\hat{J}_{0j}$ is a generator of strongly continuous one-parameter group on $(L^2_{\mathbb{C}}(\mathbb{R}^3,\varphi_a))^4$.

Exercise 1.12 Show that the spectrum (Fig. 1.9) of $\hat{J}_{03}$ is given by

$$\hbar\left(\frac{i}{2}+\mathbb{R}\right)\cup\hbar\left(-\frac{i}{2}+\mathbb{R}\right)\ .$$

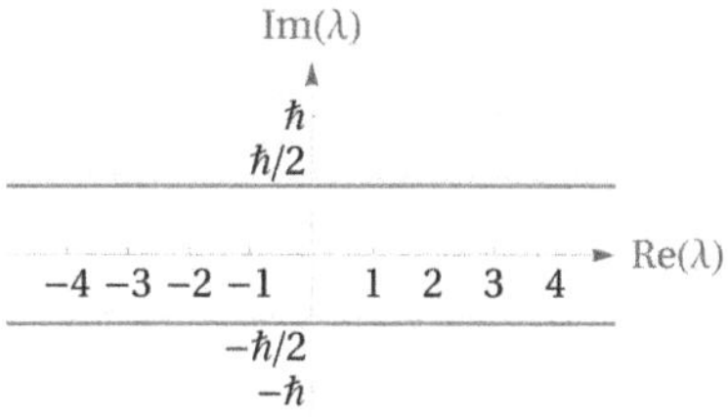

Fig. 1.9 Depiction of the spectrum (in red) of $\hat{J}_{03}$

1.10 An Extension to a Strongly Continuous Representation of $\mathbb{R}^4$ x SL(2, $\mathbb{C}$)

In the following, let $a \geqslant 0$. Using the extensions $\hat{T}_a^{\hbar_L}$ and $\hat{T}_a^{\hbar_R}$ of the Weyl representations $T_a^{\hbar_L}$ and $T_a^{\hbar_R}$, respectively, to strongly continuous representations of $\mathbb{R}^4 \times \mathrm{SL}(2, \mathbb{C})$, for $(a, G) \in \mathbb{R}^4 \times \mathrm{SL}(2, \mathbb{C})$, we define

$$(\hat{T}_a^{\hbar_L} \oplus \hat{T}_a^{\hbar_R})(a, G)f := {}^t([\hat{T}_a^{\hbar_L}(a, G)\,{}^t(f_1, f_2)]_1, [\hat{T}_a^{\hbar_L}(a, G)\,{}^t(f_1, f_2)]_2,$$
$$[\hat{T}_a^{\hbar_R}(a, G)\,{}^t(f_3, f_4)]_1, [\hat{T}_a^{\hbar_R}(a, G)\,{}^t(f_3, f_4)]_2)\ ,$$

for every $f \in (L^2_{\mathbb{C}}(\mathbb{R}^3, \varphi_a))^4$. From the linearity of $\hat{T}_a^{\hbar_L}(a, G)$ and $\hat{T}_a^{\hbar_R}(a, G)$ and the linearity of the coordinate projections of $(L^2_{\mathbb{C}}(\mathbb{R}^3, \varphi_a))^2$, it follows the linearity of $(\hat{T}_a^{\hbar_L} \oplus \hat{T}_a^{\hbar_R})(a, G)$. Further, since

$$\begin{aligned}
&\|(\hat{T}_a^{\hbar_L} \oplus \hat{T}_a^{\hbar_R})(a, G)f\|^2 \\
&= \|[\hat{T}_a^{\hbar_L}(a, G)\,{}^t(f_1, f_2)]_1\|_2^2 + \|[\hat{T}_a^{\hbar_L}(a, G)\,{}^t(f_1, f_2)]_2\|_2^2 \\
&\quad + \|[\hat{T}_a^{\hbar_R}(a, G)\,{}^t(f_3, f_4)]_1\|_2^2 + \|[\hat{T}_a^{\hbar_R}(a, G)\,{}^t(f_3, f_4)]_2\|_2^2 \\
&\leqslant 8\,\|G\|_\infty^2(\|f_1\|_2^2 + \|f_2\|_2^2) + 8\,\|G\|_\infty^2(\|f_3\|_2^2 + \|f_4\|_2^2) = 8\,\|G\|_\infty^2\|f\|^2\ ,
\end{aligned}$$

for every $f \in (L^2_{\mathbb{C}}(\mathbb{R}^3, \varphi_a))^4$, where $\|\ \|$ denotes the norm that is induced by the scalar product $\langle\,|\,\rangle$ on $(L^2_{\mathbb{C}}(\mathbb{R}^3, \varphi_a))^4$ and $\|\ \|_2$ denotes the norm that is induced by the scalar product $\langle\,|\,\rangle_2$ on $L^2_{\mathbb{C}}(\mathbb{R}^3, \varphi_a)$, it follows that $(\hat{T}_a^{\hbar_L} \oplus \hat{T}_a^{\hbar_R})(a, G) \in \mathrm{L}(X, X)$. In particular, if $G \in \mathrm{SU}(2)$, we have hat

$$\begin{aligned}
&\langle(\hat{T}_a^{\hbar_L} \oplus \hat{T}_a^{\hbar_R})(a, G)f|(\hat{T}_a^{\hbar_L} \oplus \hat{T}_a^{\hbar_R})(a, G)g\rangle \\
&= \langle[\hat{T}_a^{\hbar_L}(a, G)\,{}^t(f_1, f_2)]_1|[\hat{T}_a^{\hbar_L}(a, G)\,{}^t(g_1, g_2)]_1\rangle_2 \\
&\quad + \langle[\hat{T}_a^{\hbar_L}(a, G)\,{}^t(f_1, f_2)]_2|[\hat{T}_a^{\hbar_L}(a, G)\,{}^t(g_1, g_2)]_2\rangle_2 \\
&\quad + \langle[\hat{T}_a^{\hbar_R}(a, G)\,{}^t(f_3, f_4)]_1|[\hat{T}_a^{\hbar_R}(a, G)\,{}^t(g_3, g_4)]_1\rangle_2 \\
&\quad + \langle[\hat{T}_a^{\hbar_R}(a, G)\,{}^t(f_3, f_4)]_2|[\hat{T}_a^{\hbar_R}(a, G)\,{}^t(g_3, g_4)]_2\rangle_2 \\
&= \langle f_1|g_1\rangle_2 + \langle f_2|g_2\rangle_2 + \langle f_3|g_3\rangle_2 + \langle f_4|g_4\rangle_2 = \langle f|g\rangle \ ,
\end{aligned}$$

for all $f, g \in X$ and hence that $(\hat{T}_a^{\hbar_L} \oplus \hat{T}_a^{\hbar_R})(a, G)$ preserves the scalar product.

Further, for $(a_1, G_1), (a_2, G_2) \in \mathbb{R}^4 \times \mathrm{SL}(2, \mathbb{C})$, it follows that

$$\begin{aligned}
&(\hat{T}_a^{\hbar_L} \oplus \hat{T}_a^{\hbar_R})(a_1, G_1)(\hat{T}_a^{\hbar_L} \oplus \hat{T}_a^{\hbar_R})(a_2, G_2)f = \\
&{}^t([\hat{T}_a^{\hbar_L}(a_1, G_1)\,{}^t([\hat{T}_a^{\hbar_L}(a_2, G_2)\,{}^t(f_1, f_2)]_1, [\hat{T}_a^{\hbar_L}(a_2, G_2)\,{}^t(f_1, f_2)]_2)]_1, \\
&\quad [\hat{T}_a^{\hbar_L}(a_1, G_1)\,{}^t([\hat{T}_a^{\hbar_L}(a_2, G_2)\,{}^t(f_1, f_2)]_1, [\hat{T}_a^{\hbar_L}(a_2, G_2)\,{}^t(f_1, f_2)]_2)]_2, \\
&\quad [\hat{T}_a^{\hbar_R}(a_1, G_1)\,{}^t([\hat{T}_a^{\hbar_R}(a_2, G_2)\,{}^t(f_3, f_4)]_1, [\hat{T}_a^{\hbar_R}(a_2, G_2)\,{}^t(f_3, f_4)]_2)]_1, \\
&\quad [\hat{T}_a^{\hbar_R}(a_1, G_1)\,{}^t([\hat{T}_a^{\hbar_R}(a_2, G_2)\,{}^t(f_3, f_4)]_1, [\hat{T}_a^{\hbar_R}(a_2, G_2)\,{}^t(f_3, f_4)]_2)]_2) = \\
&{}^t([\hat{T}_a^{\hbar_L}((a_1, G_1)\cdot(a_2, G_2))\,{}^t(f_1, f_2)]_1, [\hat{T}_a^{\hbar_L}((a_1, G_1)\cdot(a_2, G_2))\,{}^t(f_1, f_2)]_2, \\
&\quad [\hat{T}_a^{\hbar_R}((a_1, G_1)\cdot(a_2, G_2))\,{}^t(f_3, f_4)]_1, [\hat{T}_a^{\hbar_R}((a_1, G_1)\cdot(a_2, G_2))\,{}^t(f_3, f_4)]_2) \\
&= (\hat{T}_a^{\hbar_L} \oplus \hat{T}_a^{\hbar_R})((a_1, G_1)\cdot(a_2, G_2))f \ ,
\end{aligned}$$

for every $f \in (L^2_{\mathbb{C}}(\mathbb{R}^3, \varphi_a))^4$ and hence that

$$(\hat{T}_a^{\hbar_L} \oplus \hat{T}_a^{\hbar_R})(a_1, G_1) \circ (\hat{T}_a^{\hbar_L} \oplus \hat{T}_a^{\hbar_R})(a_2, G_2) = (\hat{T}_a^{\hbar_L} \oplus \hat{T}_a^{\hbar_R})((a_1, G_1)\cdot(a_2, G_2)) \ .$$

Since

$$\begin{aligned}
(\hat{T}_a^{\hbar_L} \oplus \hat{T}_a^{\hbar_R})(0, E)f &= {}^t([\hat{T}_a^{\hbar_L}(0, E)\,{}^t(f_1, f_2)]_1, [\hat{T}_a^{\hbar_L}(0, E)\,{}^t(f_1, f_2)]_2, \\
&\qquad [\hat{T}_a^{\hbar_R}(0, E)\,{}^t(f_3, f_4)]_1, [\hat{T}_a^{\hbar_R}(0, E)\,{}^t(f_3, f_4)]_2) \\
&= {}^t(f_1, f_2, f_3, f_4) = f \ ,
\end{aligned}$$

for every $f \in (L^2_{\mathbb{C}}(\mathbb{R}^3, \varphi_a))^4$ and hence

$$(\hat{T}_a^{\hbar_L} \oplus \hat{T}_a^{\hbar_R})(0, E) = \mathrm{id}_{(L^2_{\mathbb{C}}(\mathbb{R}^3, \varphi_a))^4} \ ,$$

it follows that the direct sum $\hat{T}_a^{\hbar_L} \oplus \hat{T}_a^{\hbar_R}$ of the representations $\hat{T}_a^{\hbar_L}$ and $\hat{T}_a^{\hbar_R}$ is a representation of $\mathbb{R}^4 \times \mathrm{SL}(2, \mathbb{C})$. We note that this implies also that $(\hat{T}_a^{\hbar_L} \oplus \hat{T}_a^{\hbar_R})(a, G)$ is unitary for every $(a, G) \in \mathbb{R}^4 \times \mathrm{SU}(2)$. Further, if $(a, G), (\mathfrak{b}, H) \in \mathbb{R}^4 \times \mathrm{SL}(2, \mathbb{R}^4)$ and $f \in (L^2_{\mathbb{C}}(\mathbb{R}^3, \varphi_a))^4$, then

$$\begin{aligned}
&(\hat{T}_a^{\hbar_L} \oplus \hat{T}_a^{\hbar_R})(a, G)f - (\hat{T}_a^{\hbar_L} \oplus \hat{T}_a^{\hbar_R})(\mathfrak{b}, H)f \\
&= {}^t([\hat{T}_a^{\hbar_L}(a, G)\,{}^t(f_1, f_2)]_1, [\hat{T}_a^{\hbar_L}(a, G)\,{}^t(f_1, f_2)]_2, \\
&\quad [\hat{T}_a^{\hbar_R}(a, G)\,{}^t(f_3, f_4)]_1, [\hat{T}_a^{\hbar_R}(a, G)\,{}^t(f_3, f_4)]_2) - \\
&\quad {}^t([\hat{T}_a^{\hbar_L}(a, G)\,{}^t(f_1, f_2)]_1, [\hat{T}_a^{\hbar_L}(a, G)\,{}^t(f_1, f_2)]_2, \\
&\quad [\hat{T}_a^{\hbar_R}(\mathfrak{b}, H)\,{}^t(f_3, f_4)]_1, [\hat{T}_a^{\hbar_R}(\mathfrak{b}, H)\,{}^t(f_3, f_4)]_2) \\
&= {}^t([(\hat{T}_a^{\hbar_L}(a, G) - \hat{T}_a^{\hbar_L}(\mathfrak{b}, H))\,{}^t(f_1, f_2)]_1, [(\hat{T}_a^{\hbar_L}(a, G) - \hat{T}_a^{\hbar_L}(\mathfrak{b}, H))\,{}^t(f_1, f_2)]_2, \\
&\quad [(\hat{T}_a^{\hbar_R}(a, G) - \hat{T}_a^{\hbar_R}(\mathfrak{b}, H))\,{}^t(f_3, f_4)]_1, [(\hat{T}_a^{\hbar_R}(a, G) - \hat{T}_a^{\hbar_R}(\mathfrak{b}, H))\,{}^t(f_3, f_4)]_2) \,.
\end{aligned}$$

Since $\hat{T}_a^{\hbar_L}$ and $\hat{T}_a^{\hbar_R}$ are strongly continuous, it follows that if $(a_1, G_1), (a_2, G_2), \ldots$ is a sequence in $\mathbb{R}^4 \times \mathrm{SL}(2, \mathbb{C})$ that converges component-wise to $(a, G) \in \mathbb{R}^4 \times \mathrm{SL}(2, \mathbb{C})$, then

$$\lim_{\nu \to \infty} \|(\hat{T}_a^{\hbar_L} \oplus \hat{T}_a^{\hbar_R})(a_\nu, G_\nu)f - (\hat{T}_a^{\hbar_L} \oplus \hat{T}_a^{\hbar_R})(a, G)f\| = 0 \,,$$

for every $f \in (L^2_{\mathbb{C}}(\mathbb{R}^3, \varphi_a))^4$. Hence, we arrive at the following result.

If $a \geqslant 0$, then the direct sum $\hat{T}_a^{\hbar_L} \oplus \hat{T}_a^{\hbar_R}$ of the representations $\hat{T}_a^{\hbar_L}$ and $\hat{T}_a^{\hbar_R}$, defined by

$$\begin{aligned}
(\hat{T}_a^{\hbar_L} \oplus \hat{T}_a^{\hbar_R})(a, G)f := {}^t(&[\hat{T}_a^{\hbar_L}(a, G)\,{}^t(f_1, f_2)]_1, [\hat{T}_a^{\hbar_L}(a, G)\,{}^t(f_1, f_2)]_2, \\
&[\hat{T}_a^{\hbar_R}(a, G)\,{}^t(f_3, f_4)]_1, [\hat{T}_a^{\hbar_R}(a, G)\,{}^t(f_3, f_4)]_2) \,,
\end{aligned}$$

for every $f \in (L^2_{\mathbb{C}}(\mathbb{R}^3, \varphi_a))^4$ and $(a, G) \in \mathbb{R}^4 \times \mathrm{SL}(2, \mathbb{C})$, is a strongly continuous representation of $\mathbb{R}^4 \times \mathrm{SL}(2, \mathbb{C})$, i.e., a representation of $\mathbb{R}^4 \times \mathrm{SL}(2, \mathbb{C})$ such that if $a_1, a_2, \ldots$ is a sequence in $\mathbb{R}^4$ that is component-wise convergent to $a \in \mathbb{R}^4$ and $G_1, G_2, \ldots$ is a sequence in $\mathrm{SL}(2, \mathbb{C})$ that converges component-wise to $G \in \mathrm{SL}(2, \mathbb{C})$, then

$$\lim_{\nu \to \infty} \|(\hat{T}_a^{\hbar_L} \oplus \hat{T}_a^{\hbar_R})(a_\nu, G_\nu)f - (\hat{T}_a^{\hbar_L} \oplus \hat{T}_a^{\hbar_R})(a, G)f\| = 0 \,,$$

for every $f \in (L^2_{\mathbb{C}}(\mathbb{R}^3, \varphi_a))^2$, where $\| \; \|$ denotes the norm that is induced by the scalar product $\langle \, | \, \rangle$ on $(L^2_{\mathbb{C}}(\mathbb{R}^3, \varphi_a))^4$. Further, $(\hat{T}^{\hbar_L}_a \oplus \hat{T}^{\hbar_R}_a)(a, G)$ is unitary for every $(a, G) \in \mathbb{R}^4 \times \mathrm{SU}(2)$.

1.10.1 Generators Associated with Translations

For $a \in \mathbb{R}^4$, we define $M : \mathbb{R} \to \mathbb{R}^4 \times \mathrm{SL}(2, \mathbb{C})$, by

$$M(s) := (sa, E_{2\times 2}) \ ,$$

for every $s \in \mathbb{R}$. According to Sect. 1.6.1, M is a component-wise continuous one-parameter subgroup of $\mathbb{R}^4 \times \mathrm{SU}(2)$. Further, according to (1.24), for $\hbar \in \{\hbar_R, \hbar_L\}$, every $s \in \mathbb{R}$ and $f \in (L^2_{\mathbb{C}}(\mathbb{R}^3, \varphi_a))^2$, we have that

$$\hat{T}^{\hbar}_a(M(s)) \begin{pmatrix} f_1 \\ f_2 \end{pmatrix} = \begin{pmatrix} \exp(is\, T_{a_0 v^0_a - \vec{a}\cdot \mathrm{id}_{\mathbb{R}^3}}) f_1 \\ \exp(is\, T_{a_0 v^0_a - \vec{a}\cdot \mathrm{id}_{\mathbb{R}^3}}) f_2 \end{pmatrix} \ ,$$

where $T_{a_0 v^0_a - \vec{a}\cdot \mathrm{id}_{\mathbb{R}^3}}$ denotes the maximal multiplication operator in $L^2_{\mathbb{C}}(\mathbb{R}^3, \varphi_a)$ corresponding to the function $a_0 v^0_a - \vec{a} \cdot \mathrm{id}_{\mathbb{R}^3}$, $\vec{a} := {}^t(a_1, a_2, a_3)$ $\vec{a} := {}^t(a_1, a_2, a_3)$, $\vec{a} \cdot \mathrm{id}_{\mathbb{R}^3} : \mathbb{R}^3 \to \mathbb{R}^3$ is defined by $(\vec{a} \cdot \mathrm{id}_{\mathbb{R}^3})(v) := \vec{a} \cdot v$, for every $v \in \mathbb{R}^3$. Hence,

$$\begin{aligned}
&(\hat{T}^{\hbar_L}_a \oplus \hat{T}^{\hbar_R}_a)(M(s)) f \\
&= {}^t([\hat{T}^{\hbar_L}_a(M(s))\, {}^t(f_1, f_2)]_1, [\hat{T}^{\hbar_L}_a(M(s))\, {}^t(f_1, f_2)]_2, \\
&\qquad [\hat{T}^{\hbar_R}_a(M(s))\, {}^t(f_3, f_4)]_1, [\hat{T}^{\hbar_R}_a(M(s))\, {}^t(f_3, f_4)]_2) \\
&= {}^t(\exp(is\, T_{a_0 v^0_a - \vec{a}\cdot \mathrm{id}_{\mathbb{R}^3}}) f_1, \exp(is\, T_{a_0 v^0_a - \vec{a}\cdot \mathrm{id}_{\mathbb{R}^3}}) f_2, \\
&\qquad \exp(is\, T_{a_0 v^0_a - \vec{a}\cdot \mathrm{id}_{\mathbb{R}^3}}) f_3, \exp(is\, T_{a_0 v^0_a - \vec{a}\cdot \mathrm{id}_{\mathbb{R}^3}}) f_4) \ ,
\end{aligned}$$

for every $f \in (L^2_{\mathbb{C}}(\mathbb{R}^3, \varphi_a))^4$. Therefore, $(\hat{T}^{\hbar_L}_a \oplus \hat{T}^{\hbar_R}_a) \circ M$ is a strongly continuous one-parameter unitary group. According to Stone's theorem, there is a unique densely-defined, linear and self-adjoint operator A_M in $X := (L^2_{\mathbb{C}}(\mathbb{R}^3, \varphi_a))^4$ such that

$$\exp(isA_M) = ((\hat{T}^{\hbar_L}_a \oplus \hat{T}^{\hbar_R}_a) \circ M)(s) \ ,$$

for every $s \in \mathbb{R}$ and, in particular, that $A_M : D(A_M) \to X$ is given by

$$D(A_M) = \{f \in X : \lim_{s \to 0, s \neq 0} \frac{1}{s} \left[((\hat{T}^{\hbar_L}_a \oplus \hat{T}^{\hbar_R}_a) \circ M)(s) - \mathrm{id}_X\right] f \text{ exists}\}$$

and for every $f \in D(A_M)$

$$A_M f = \frac{1}{i} \lim_{s \to 0, s \neq 0} \frac{1}{s} \left[((\hat{T}_a^{\hbar_L} \oplus \hat{T}_a^{\hbar_R}) \circ M)(s) - \mathrm{id}_X \right] f \,.$$

Further, for $f \in X$ and $s \in \mathbb{R}^*$, it follows that

$$\frac{1}{is} \left[((\hat{T}_a^{\hbar_L} \oplus \hat{T}_a^{\hbar_R}) \circ M)(s) - \mathrm{id}_X \right] f = \begin{pmatrix} \frac{1}{is} \, [\exp(is \, T_{a_0 v_a^0 - \vec{a} \cdot \mathrm{id}_{\mathbb{R}^3}}) - 1] f_1 \\ \frac{1}{is} \, [\exp(is \, T_{a_0 v_a^0 - \vec{a} \cdot \mathrm{id}_{\mathbb{R}^3}}) - 1] f_2 \\ \frac{1}{is} \, [\exp(is \, T_{a_0 v_a^0 - \vec{a} \cdot \mathrm{id}_{\mathbb{R}^3}}) - 1] f_3 \\ \frac{1}{is} \, [\exp(is \, T_{a_0 v_a^0 - \vec{a} \cdot \mathrm{id}_{\mathbb{R}^3}}) - 1] f_4 \end{pmatrix} .$$

We note that the coordinate projections $p_k : X \to L^2_{\mathbb{C}}(\mathbb{R}^3, \varphi_a)$ as well as the inclusions $\iota_k : L^2_{\mathbb{C}}(\mathbb{R}^3, \varphi_a) \hookrightarrow X, k \in \{1, 2, 3, 4\}$, defined by $p_k f := f_k$, for every $f \in X$ and $\iota_1 f := {}^t(f, 0, 0, 0)$, $\iota_2 f := {}^t(0, f, 0, 0)$, $\iota_3 f := {}^t(0, 0, f, 0)$ and $\iota_4 f := {}^t(0, 0, 0, f)$, for every $f \in L^2_{\mathbb{C}}(\mathbb{R}^3, \varphi_a)$ are linear and continuous, since

$$\|p_k f\|_2 = \|f_k\|_2 \leqslant \|f\| \;, \;\; \|\iota_k g\| = \|g\|_2 \;,$$

for every $f \in X, g \in L^2_{\mathbb{C}}(\mathbb{R}^3, \varphi_a)$ and $k \in \{1, 2, 3, 4\}$. Hence, $f \in D(A_M)$ if and only if f is part of the domain of

$$\bigtimes_{k=1}^{4} T_{a_0 v_a^0 - \vec{a} \cdot \mathrm{id}_{\mathbb{R}^3}}$$

and if $f \in D(A_M)$, then

$$A_M f = \left[\bigtimes_{k=1}^{4} T_{a_0 v_a^0 - \vec{a} \cdot \mathrm{id}_{\mathbb{R}^3}} \right] f \,.$$

Hence, we arrive at the following result.

Generators Associated with Translations in $(L^2_{\mathbb{C}}(\mathbb{R}^3, \varphi_a))^4$ For $a \in \mathbb{R}^4$, we have that

$$((\hat{T}_a^{\hbar_L} \oplus \hat{T}_a^{\hbar_R}) \circ M)(s) = \exp\left(is \bigtimes_{k=1}^{4} T_{a_0 v_a^0 - \vec{a} \cdot \mathrm{id}_{\mathbb{R}^3}} \right) ,$$

for every $s \in \mathbb{R}$, where the one-parameter subgroup of $\mathbb{R}^4 \times \mathrm{SL}(2, \mathbb{C})$, $M : \mathbb{R} \to \mathbb{R}^4 \times \mathrm{SL}(2, \mathbb{C})$ is defined by $M(s) := (sa, E)$, for every $s \in \mathbb{R}$, $T_{a_0 v_a^0 - \vec{a} \cdot \mathrm{id}_{\mathbb{R}^3}}$ denotes the maximal multiplication operator in $L^2_{\mathbb{C}}(\mathbb{R}^3, \varphi_a)$, corresponding to the function $a_0 v_a^0 - \vec{a} \cdot \mathrm{id}_{\mathbb{R}^3}$, $\vec{a} := {}^t(a_1, a_2, a_3)$ and $\vec{a} \cdot \mathrm{id}_{\mathbb{R}^3} : \mathbb{R}^3 \to \mathbb{R}^3$ is defined by $(\vec{a} \cdot \mathrm{id}_{\mathbb{R}^3})(v) := \vec{a} \cdot v$, for every $v \in \mathbb{R}^3$.

Exercise 1.13 Show that the spectrum of

$$\bigtimes_{k=1}^{4} T_{a_0 v_a^0 - \vec{a} \cdot \mathrm{id}_{\mathbb{R}^3}}$$

is given by the closure of the range of $a_0 v_a^0 - \vec{a} \cdot \mathrm{id}_{\mathbb{R}^3}$

$$\overline{\mathrm{Ran}(a_0 v_a^0 - \vec{a} \cdot \mathrm{id}_{\mathbb{R}^3})} \ .$$

Exercise 1.14 Sow that, if $a \in \mathbb{R}^4$ is future-oriented, timelike, and, in particular, such that $a \cdot a = 1$, the spectrum of $\bigtimes_{k=1}^{4} T_{a_0 v_a^0 - \vec{a} \cdot \mathrm{id}_{\mathbb{R}^3}}$ is given by the closed interval $[a, \infty)$.

As a side remark, using the Hilbert space isomorphism $V_a : L^2_{\mathbb{C}}(\mathbb{R}^3, \varphi_a) \to L^2_{\mathbb{C}}(\mathbb{R}^3)$ from Exercise 5.17 of Part I, defined by

$$V_a f := (v_a^0)^{-1/2} f \ ,$$

for every $f \in L^2_{\mathbb{C}}(\mathbb{R}^3, \varphi_a)$, with inverse $V_a^{-1} : L^2_{\mathbb{C}}(\mathbb{R}^3) \to L^2_{\mathbb{C}}(\mathbb{R}^3, \varphi_a)$, given by

$$V_a^{-1} f := (v_a^0)^{1/2} f \ ,$$

for every $f \in L^2_{\mathbb{C}}(\mathbb{R}^3)$, we note that

$$\bigtimes_{k=1}^{4} V_a : (L^2_{\mathbb{C}}(\mathbb{R}^3, \varphi_a))^4 \to (L^2_{\mathbb{C}}(\mathbb{R}^3))^4$$

is a Hilbert space isomorphism and that

$$\left(\bigtimes_{k=1}^{4} V_a\right) ((\hat{T}_a^{\hbar_L} \oplus \hat{T}_a^{\hbar_R})(M(s))) \left(\bigtimes_{k=1}^{4} V_a\right)^{-1} f$$
$$= \exp\left(is \bigtimes_{k=1}^{4} T_{a_0 v_a^0 - \vec{a} \cdot \mathrm{id}_{\mathbb{R}^3}}\right) f \ ,$$

for every $s \in \mathbb{R}$ and $f \in (L^2_{\mathbb{C}}(\mathbb{R}^3))^4$, where $T_{a_0 v_a^0 - \vec{a} \cdot \mathrm{id}_{\mathbb{R}^3}}$ denotes the maximal multiplication operator in $L^2_{\mathbb{C}}(\mathbb{R}^3)$, corresponding to the function $a_0 v_a^0 - \vec{a} \cdot \mathrm{id}_{\mathbb{R}^3}$, $\vec{a} := {}^t(a_1, a_2, a_3)$. As a consequence, we also have the following.

Generators Associated with Translations in $(L^2_{\mathbb{C}}(\mathbb{R}^3))^2$ For $a \in \mathbb{R}^4$, we have that

$$\left(\bigtimes_{k=1}^{4} V_a\right)((\hat{T}_a^{\hbar_L} \oplus \hat{T}_a^{\hbar_R})(M(s))\left(\bigtimes_{k=1}^{4} V_a\right)^{-1} f$$
$$= \exp\left(is \bigtimes_{k=1}^{4} T_{a_0 v_a^0 - \vec{a} \cdot \mathrm{id}_{\mathbb{R}^3}}\right) f \,,$$

for every $s \in \mathbb{R}$, where the one-parameter subgroup of $\mathbb{R}^4 \times \mathrm{SL}(2, \mathbb{C})$, $M : \mathbb{R} \to \mathbb{R}^4 \times \mathrm{SL}(2, \mathbb{C})$ is defined by $M(s) := (sa, E)$, for every $s \in \mathbb{R}$, $T_{a_0 v_a^0 - \vec{a} \cdot \mathrm{id}_{\mathbb{R}^3}}$ denotes the maximal multiplication operator in $L^2_{\mathbb{C}}(\mathbb{R}^3)$, corresponding to the function $a_0 v_a^0 - \vec{a} \cdot \mathrm{id}_{\mathbb{R}^3}$, $\vec{a} := {}^t(a_1, a_2, a_3)$ and $\vec{a} \cdot \mathrm{id}_{\mathbb{R}^3} : \mathbb{R}^3 \to \mathbb{R}^3$ is defined by $(\vec{a} \cdot \mathrm{id}_{\mathbb{R}^3})(v) := \vec{a} \cdot v$, for every $v \in \mathbb{R}^3$.

1.11 Dirac Equation

In the following, we are going to derive the Dirac equation from the Weyl equations, a derivation that is attributed to van der Waerden [64]. If $\psi_L, \psi_R \in (C^1(\mathbb{R}^4, \mathbb{C}))^2$ are solutions of the Weyl equations

$$\left(\frac{\partial}{\partial u_0} - \sum_{\alpha=1}^{3} \sigma_\alpha \frac{\partial}{\partial u_\alpha}\right)\psi_L = 0 \,, \quad \left(\frac{\partial}{\partial u_0} + \sum_{\alpha=1}^{3} \sigma_\alpha \frac{\partial}{\partial u_\alpha}\right)\psi_R = 0 \,,$$

we obtain the system of equations

$$\frac{\partial \psi}{\partial u_0} + \sum_{\alpha=1}^{3} \begin{pmatrix} -(\sigma_\alpha)_{11} & -(\sigma_\alpha)_{12} & 0 & 0 \\ -(\sigma_\alpha)_{21} & -(\sigma_\alpha)_{22} & 0 & 0 \\ 0 & 0 & (\sigma_\alpha)_{11} & (\sigma_\alpha)_{12} \\ 0 & 0 & (\sigma_\alpha)_{21} & (\sigma_\alpha)_{22} \end{pmatrix} \cdot \frac{\partial \psi}{\partial u_\alpha} = 0 \,,$$

where $\psi \in (C^1(\mathbb{R}^4, \mathbb{C}))^4$ is defined by

$$\psi := \begin{pmatrix} (\psi_L)_1 \\ (\psi_L)_2 \\ (\psi_R)_1 \\ (\psi_R)_2 \end{pmatrix} .$$

Multiplication of this system from the left by the involutory matrix

$$\gamma_0 := \begin{pmatrix} 0 & 0 & 1 & 0 \\ 0 & 0 & 0 & 1 \\ 1 & 0 & 0 & 0 \\ 0 & 1 & 0 & 0 \end{pmatrix}$$

leads to the equivalent system

$$i\hbar \sum_{k=0}^{3} \gamma_k \cdot \frac{\partial \psi}{\partial u_k} = 0 \ ,$$

the Dirac equation, describing the propagation of massless field Spin $1/2$ field in Minkowski space-time, where

$$\begin{aligned}
\gamma_\alpha := \gamma_0 \cdot &\begin{pmatrix} -(\sigma_\alpha)_{11} & -(\sigma_\alpha)_{12} & 0 & 0 \\ -(\sigma_\alpha)_{21} & -(\sigma_\alpha)_{22} & 0 & 0 \\ 0 & 0 & (\sigma_\alpha)_{11} & (\sigma_\alpha)_{12} \\ 0 & 0 & (\sigma_\alpha)_{21} & (\sigma_\alpha)_{22} \end{pmatrix} \\
= &\begin{pmatrix} 0 & 0 & (\sigma_\alpha)_{11} & (\sigma_\alpha)_{12} \\ 0 & 0 & (\sigma_\alpha)_{21} & (\sigma_\alpha)_{22} \\ -(\sigma_\alpha)_{11} & -(\sigma_\alpha)_{12} & 0 & 0 \\ -(\sigma_\alpha)_{21} & -(\sigma_\alpha)_{22} & 0 & 0 \end{pmatrix} \\
= &\begin{pmatrix} (\sigma_\alpha)_{11} & (\sigma_\alpha)_{12} & 0 & 0 \\ (\sigma_\alpha)_{21} & (\sigma_\alpha)_{22} & 0 & 0 \\ 0 & 0 & -(\sigma_\alpha)_{11} & -(\sigma_\alpha)_{12} \\ 0 & 0 & -(\sigma_\alpha)_{21} & -(\sigma_\alpha)_{22} \end{pmatrix} \cdot \begin{pmatrix} 0 & 0 & 1 & 0 \\ 0 & 0 & 0 & 1 \\ 1 & 0 & 0 & 0 \\ 0 & 1 & 0 & 0 \end{pmatrix} \ , \\
= -&\begin{pmatrix} -(\sigma_\alpha)_{11} & -(\sigma_\alpha)_{12} & 0 & 0 \\ -(\sigma_\alpha)_{21} & -(\sigma_\alpha)_{22} & 0 & 0 \\ 0 & 0 & (\sigma_\alpha)_{11} & (\sigma_\alpha)_{12} \\ 0 & 0 & (\sigma_\alpha)_{21} & (\sigma_\alpha)_{22} \end{pmatrix} \cdot \gamma_0 \ ,
\end{aligned}$$

for every $\alpha \in \{1, 2, 3\}$ and $\hbar$ is the reduced Planck's constant. We note that

$$\gamma_0 \cdot \gamma_\alpha = -\gamma_\alpha \cdot \gamma_0 \ ,$$

for $\alpha \in \{1, 2, 3\}$. Further, for $\alpha, \beta \in \{1, 2, 3\}$, we have that

$$\begin{aligned}
\sigma_\alpha \cdot \sigma_\beta &= \begin{pmatrix} (\sigma_\alpha)_{11} & (\sigma_\alpha)_{12} \\ (\sigma_\alpha)_{21} & (\sigma_\alpha)_{22} \end{pmatrix} \cdot \begin{pmatrix} (\sigma_\beta)_{11} & (\sigma_\beta)_{12} \\ (\sigma_\beta)_{21} & (\sigma_\beta)_{22} \end{pmatrix} \\
&= \begin{pmatrix} (\sigma_\alpha)_{11}(\sigma_\beta)_{11} + (\sigma_\alpha)_{12}(\sigma_\beta)_{21} & (\sigma_\alpha)_{11}(\sigma_\beta)_{12} + (\sigma_\alpha)_{12}(\sigma_\beta)_{22} \\ (\sigma_\alpha)_{21}(\sigma_\beta)_{11} + (\sigma_\alpha)_{22}(\sigma_\beta)_{21} & (\sigma_\alpha)_{21}(\sigma_\beta)_{12} + (\sigma_\alpha)_{22}(\sigma_\beta)_{22} \end{pmatrix}
\end{aligned}$$

and hence that

$$
\begin{aligned}
&\gamma_\alpha \cdot \gamma_\beta \\
&= \begin{pmatrix} 0 & 0 & (\sigma_\alpha)_{11} & (\sigma_\alpha)_{12} \\ 0 & 0 & (\sigma_\alpha)_{21} & (\sigma_\alpha)_{22} \\ -(\sigma_\alpha)_{11} & -(\sigma_\alpha)_{12} & 0 & 0 \\ -(\sigma_\alpha)_{21} & -(\sigma_\alpha)_{22} & 0 & 0 \end{pmatrix} \cdot \begin{pmatrix} 0 & 0 & (\sigma_\beta)_{11} & (\sigma_\beta)_{12} \\ 0 & 0 & (\sigma_\beta)_{21} & (\sigma_\beta)_{22} \\ -(\sigma_\beta)_{11} & -(\sigma_\beta)_{12} & 0 & 0 \\ -(\sigma_\beta)_{21} & -(\sigma_\beta)_{22} & 0 & 0 \end{pmatrix} \\
&= - \begin{pmatrix} (\sigma_\alpha \cdot \sigma_\beta)_{11} & (\sigma_\alpha \cdot \sigma_\beta)_{12} & 0 & 0 \\ (\sigma_\alpha \cdot \sigma_\beta)_{21} & (\sigma_\alpha \cdot \sigma_\beta)_{22} & 0 & 0 \\ 0 & 0 & (\sigma_\alpha \cdot \sigma_\beta)_{11} & (\sigma_\alpha \cdot \sigma_\beta)_{12} \\ 0 & 0 & (\sigma_\alpha \cdot \sigma_\beta)_{21} & (\sigma_\alpha \cdot \sigma_\beta)_{22} \end{pmatrix} .
\end{aligned}
$$

As a consequence,

$$\gamma_\alpha \cdot \gamma_\alpha = -E \ , \ \gamma_\alpha \cdot \gamma_\beta = -\gamma_\beta \cdot \gamma_\alpha \ ,$$

for $\alpha, \beta \in \{1, 2, 3\}$ such that $\beta \neq \alpha$, where we use the multiplication table of the Pauli spin matrices, Table 4.1, from Part I. Hence, it follows that

$$\boxed{\gamma_k \gamma_l + \gamma_l \gamma_k = 2\eta_{kl} E \ ,}$$

for all $k, l \in \{0, 1, 2, 3\}$, where

$$\eta := \begin{pmatrix} 1 & 0 & 0 & 0 \\ 0 & -1 & 0 & 0 \\ 0 & 0 & -1 & 0 \\ 0 & 0 & 0 & -1 \end{pmatrix} .$$

The representation of the Dirac matrices $\gamma_0, \gamma_1, \gamma_2, \gamma_3$ thus obtained is called the chiral representation or the Weyl representation. We note that for $f \in (C^2(\mathbb{R}^4, \mathbb{C}))^4$, we have that

$$
\begin{aligned}
&\left(i\hbar \sum_{k=0}^{3} \gamma_k \frac{\partial}{\partial u_k} \right) \left(i\hbar \sum_{l=0}^{3} \gamma_l \frac{\partial f}{\partial u_l} \right) = -\hbar^2 \sum_{k,l=0}^{3} \gamma_k \gamma_l \frac{\partial^2 f}{\partial u_k \partial u_l} \\
&= -\frac{\hbar^2}{2} \sum_{k,l=0}^{3} (\gamma_k \gamma_l + \gamma_l \gamma_k) \frac{\partial^2 f}{\partial u_k \partial u_l} = -\hbar^2 \sum_{k,l=0}^{3} \eta_{kl} \frac{\partial^2 f}{\partial u_k \partial u_l} = -\hbar^2 \Box f \ ,
\end{aligned}
$$

where $\Box$ denotes the d'Alembertian operator. Hence, if $\psi \in (C^2(\mathbb{R}^4, \mathbb{C}))^4$ satisfies the Dirac equation,

$$i\hbar \sum_{k=0}^{3} \gamma_k \cdot \frac{\partial \psi}{\partial u_k} - \frac{mc}{\kappa} \psi = 0 \ , \tag{1.28}$$

describing the propagation of a Spin 1/2 field with mass $m \geqslant 0$ in Minkowski space-time, where c denotes the speed of light in vacuum and $\kappa > 0$ is a scale factor with dimension 1/length, then

$$\begin{aligned}
0 &= i\hbar \sum_{k=0}^{3} \gamma_k \cdot \frac{\partial}{\partial u_k}\left(i\hbar \sum_{l=0}^{3} \gamma_l \cdot \frac{\partial \psi}{\partial u_l} - \frac{mc}{\kappa}\,\psi \right) \\
&= -\hbar^2\,\Box\psi - \frac{mc}{\kappa}\, i\hbar \sum_{k=0}^{3} \gamma_k \frac{\partial \psi}{\partial u_k} = -\hbar^2\,\Box\psi - \frac{mc}{\kappa}\frac{mc}{\kappa}\,\psi \\
&= -\hbar^2 \left[\Box + \left(\frac{mc}{\hbar\kappa}\right)^2 \right] \psi\,,
\end{aligned}$$

i.e., every component of ψ satisfies the Klein-Gordon equation, the latter describing the propagation of a Spin 0 field with mass m in Minkowski space-time.

As is the case for the Weyl equations, to every $(a, G) \in \mathbb{R}^4 \times \mathrm{SL}(2, \mathbb{C})$, there corresponds a symmetry transformation of the Dirac equation, i.e., a transformation that maps solutions of the equation into such solutions. Indeed, if $\psi \in (C^1(\mathbb{R}^4, \mathbb{C}))^4$ is a solution of the Dirac equation (1.28) and $\psi_L, \psi_R \in (C^1(\mathbb{R}^4, \mathbb{C}))^2$ are defined by

$$\psi_L := \begin{pmatrix} \psi_1 \\ \psi_2 \end{pmatrix}\,, \quad \psi_R := \begin{pmatrix} \psi_3 \\ \psi_4 \end{pmatrix}\,,$$

it follows that

$$\begin{aligned}
\left(\frac{\partial}{\partial u_0} - \sum_{\alpha=1}^{3} \sigma_\alpha \frac{\partial}{\partial u_\alpha} \right) \psi_L &= -i\,\frac{mc}{\hbar\kappa}\,\psi_R\,, \\
\left(\frac{\partial}{\partial u_0} + \sum_{\alpha=1}^{3} \sigma_\alpha \frac{\partial}{\partial u_\alpha} \right) \psi_R &= -i\,\frac{mc}{\hbar\kappa}\,\psi_L\,,
\end{aligned} \tag{1.29}$$

and hence, using the results of Sect. 1.7, for $(a, G) \in \mathbb{R}^4 \times \mathrm{SL}(2, \mathbb{C})$, that

$$\begin{aligned}
\sum_{k=0}^{3} \bar{\sigma}_k \cdot \frac{\partial R_{\hbar_L}((a,G))\psi_L}{\partial u_k} &= [R_{\hbar_R}((a,G))]\Big(\sum_{k=0}^{3} \bar{\sigma}_k \cdot \frac{\partial \psi_L}{\partial u_l} \Big) \\
&= -i\,\frac{mc}{\hbar\kappa}\,[R_{\hbar_R}((a,G))]\psi_R\,, \\
\sum_{k=0}^{3} \sigma_k \cdot \frac{\partial R_{\hbar_R}((a,G))\psi_R}{\partial u_k} &= [R_{\hbar_L}((a,G))]\Big(\sum_{k=0}^{3} \sigma_k \cdot \frac{\partial \psi_R}{\partial u_l} \Big) \\
&= -i\,\frac{mc}{\hbar\kappa}\,[R_{\hbar_L}((a,G))]\psi_L\,.
\end{aligned}$$

The latter implies that

$$\left(i\hbar\sum_{k=0}^{3}\gamma_k\cdot\frac{\partial}{\partial u_k}-\frac{mc}{\kappa}\right)\begin{pmatrix}(R_{\hbar_L}((a,G))\psi_L)_1\\(R_{\hbar_L}((a,G))\psi_L)_2\\(R_{\hbar_R}((a,G))\psi_R)_1\\(R_{\hbar_R}((a,G))\psi_R)_2\end{pmatrix}=0\ ,$$

Hence, we obtain the following result.

if $\psi\in(C^1(\mathbb{R}^4,\mathbb{C}))^4$ is a solution of the Dirac equation, i.e.,

$$i\hbar\sum_{k=0}^{3}\gamma_k\cdot\frac{\partial\psi}{\partial u_k}-\frac{mc}{\kappa}\psi=0\ ,$$

and $\psi_L,\psi_R\in(C^1(\mathbb{R}^4,\mathbb{C}))^2$ are defined by

$$\psi_L:=\begin{pmatrix}\psi_1\\\psi_2\end{pmatrix}\ ,\ \psi_R:=\begin{pmatrix}\psi_3\\\psi_4\end{pmatrix}\ ,$$

then

$$\left(i\hbar\sum_{k=0}^{3}\gamma_k\cdot\frac{\partial}{\partial u_k}-\frac{mc}{\kappa}\right)\begin{pmatrix}(R_{\hbar_L}((a,G))\psi_L)_1\\(R_{\hbar_L}((a,G))\psi_L)_2\\(R_{\hbar_R}((a,G))\psi_R)_1\\(R_{\hbar_R}((a,G))\psi_R)_2\end{pmatrix}=0\ ,$$

where for $\hbar\in\{\hbar_R,\hbar_L\}$

$$[R_\hbar((a,G))f](u):=\hbar(G)\cdot f([\Phi_2(G)]^{-1}\cdot(u-a))\ ,$$

for every $f\in(C^1(\mathbb{R}^4,\mathbb{C}))^2$ and $u\in\mathbb{R}^4$.

Exercise 1.15 We denote by $\mathbb{C}^{\mathbb{R}^4}$ the complex vector space of all maps from $\mathbb{R}^4$ to $\mathbb{C}$. Further, for every $(a,G)\in\mathbb{R}^4\times\mathrm{SL}(2,\mathbb{C})$, we define

$$R((a,G))f:=\begin{pmatrix}(R_{\hbar_L}((a,G))^t(f_1,f_2))_1\\(R_{\hbar_L}((a,G))^t(f_1,f_2))_2\\(R_{\hbar_R}((a,G))^t(f_3,f_4))_1\\(R_{\hbar_R}((a,G))^t(f_3,f_4))_2\end{pmatrix}\ ,$$

for every $f\in(\mathbb{C}^{\mathbb{R}^4})^4$. Show that $R((a,G))\in\mathrm{Hom}((\mathbb{C}^{\mathbb{R}^4})^4,(\mathbb{C}^{\mathbb{R}^4})^4)$, for every $(a,G)\in\mathbb{R}^4\times\mathrm{SL}(2,\mathbb{C})$. Further, show that the map $R:\mathbb{R}^4\times\mathrm{SL}(2,\mathbb{C})\to\mathrm{Hom}((\mathbb{C}^{\mathbb{R}^4})^4,(\mathbb{C}^{\mathbb{R}^4})^4)$, that associates with every $(a, G) \in \mathbb{R}^4 \times \mathrm{SL}(2, \mathbb{C})$ the map $R((a, G))$, is a representation of $\mathbb{R}^4 \times \mathrm{SL}(2, \mathbb{C})$ that leaves $(C^1(\mathbb{R}^4, \mathbb{C}))^4$ invariant.

In the following, we represent the Dirac equation, in the form (1.29), in form of a Schrödinger equation in $(L^2_{\mathbb{C}}(\mathbb{R}^3, \varphi_a))^4$, where $a \geqslant 0$. For this purpose, we consider the linear operator

$$A_D : (D(T_{v_1}) \cap D(T_{v_2}) \cap D(T_{v_3}))^4 \to (L^2_{\mathbb{C}}(\mathbb{R}^3, \varphi_a))^4 \,,$$

defined by

$$\begin{aligned}
A_D f &:= \sum_{\alpha=1}^{3} \begin{pmatrix} -(\sigma_\alpha)_{11} & -(\sigma_\alpha)_{12} & 0 & 0 \\ -(\sigma_\alpha)_{21} & -(\sigma_\alpha)_{22} & 0 & 0 \\ 0 & 0 & (\sigma_\alpha)_{11} & (\sigma_\alpha)_{12} \\ 0 & 0 & (\sigma_\alpha)_{21} & (\sigma_\alpha)_{22} \end{pmatrix} \cdot \begin{pmatrix} T_{v_\alpha} f_1 \\ T_{v_\alpha} f_2 \\ T_{v_\alpha} f_3 \\ T_{v_\alpha} f_4 \end{pmatrix} \\
&\quad + \frac{mc}{\hbar\kappa} \begin{pmatrix} 0 & 0 & 1 & 0 \\ 0 & 0 & 0 & 1 \\ 1 & 0 & 0 & 0 \\ 0 & 1 & 0 & 0 \end{pmatrix} \cdot \begin{pmatrix} f_1 \\ f_2 \\ f_3 \\ f_4 \end{pmatrix} \\
&= \begin{pmatrix} -T_{v_3} f_1 - (T_{v_1} - i T_{v_2}) f_2 + \mu f_3 \\ -(T_{v_1} + i T_{v_2}) f_1 + T_{v_3} f_2 + \mu f_4 \\ \mu f_1 + T_{v_3} f_3 + (T_{v_1} - i T_{v_2}) f_4 \\ \mu f_2 + (T_{v_1} + i T_{v_2}) f_3 - T_{v_3} f_4 \end{pmatrix} \\
&= \begin{pmatrix} -v_3 & -v_1 + i\, v_2 & \mu & 0 \\ -v_1 - i\, v_2 & v_3 & 0 & \mu \\ \mu & 0 & v_3 & v_1 - i\, v_2 \\ 0 & \mu & v_1 + i\, v_2 & -v_3 \end{pmatrix} \cdot \begin{pmatrix} f_1 \\ f_2 \\ f_3 \\ f_4 \end{pmatrix}
\end{aligned}$$

for every $f \in (D(T_{v_1}) \cap D(T_{v_2}) \cap D(T_{v_3}))^4$, where $\mu := mc/(\hbar\kappa) \geqslant 0$, for $\alpha \in \{1, 2, 3\}$, $v_\alpha : \mathbb{R}^3 \to \mathbb{R}$ is the coordinate projection of $\mathbb{R}^3$ onto the α-th component and T_{v_α} denotes the maximal multiplication operator in $L^2_{\mathbb{C}}(\mathbb{R}^3, \varphi_a)$ with v_α, with domain $D(T_{v_\alpha})$. Associated with the Dirac equation is the Hamilton operator

$$H_D = \hbar\kappa c A_D \,.$$

We note that A_D is densely-defined, since $(C_0(\mathbb{R}^3, \mathbb{C}))^4 \subset (D(T_{v_1}) \cap D(T_{v_2}) \cap D(T_{v_3}))^4$. Further, A_D is symmetric, since we have that

$$
\begin{aligned}
\langle f|A_D g\rangle &= \langle f_1| - T_{v_3}g_1 - (T_{v_1} - iT_{v_2})g_2 + \mu g_3\rangle_2 \\
&\quad + \langle f_2| - (T_{v_1} + iT_{v_2})g_1 + T_{v_3}g_2 + \mu g_4\rangle_2 \\
&\quad + \langle f_3|\mu g_1 + T_{v_3}g_3 + (T_{v_1} - iT_{v_2})g_4\rangle_2 \\
&\quad + \langle f_4|\mu g_2 + (T_{v_1} + iT_{v_2})g_3 - T_{v_3}g_4\rangle_2 \\
&= \langle -T_{v_3}f_1|g_1\rangle_2 + \langle -(T_{v_1} + iT_{v_2})f_1|g_2\rangle + \langle \mu f_1|g_3\rangle_2 \\
&\quad + \langle -(T_{v_1} - iT_{v_2})f_2|g_1\rangle_2 + \langle T_{v_3}f_2|g_2\rangle_2 + \langle \mu f_2|g_4\rangle_2 \\
&\quad + \langle \mu f_3|g_1\rangle_2 + \langle T_{v_3}f_3|g_3\rangle_2 + \langle (T_{v_1} + iT_{v_2})f_3|g_4\rangle_2 \\
&\quad + \langle \mu f_4|g_2\rangle_2 + \langle (T_{v_1} - iT_{v_2})f_4|g_3\rangle_2 + \langle -T_{v_3}f_4|g_4\rangle_2 \\
&= \langle -T_{v_3}f_1 - (T_{v_1} - iT_{v_2})f_2 + \mu f_3|g_1\rangle_2 \\
&\quad + \langle -(T_{v_1} + iT_{v_2})f_1 + T_{v_3}f_2 + \mu f_4|g_2\rangle_2 \\
&\quad + \langle \mu f_1 + T_{v_3}f_3 + (T_{v_1} - iT_{v_2})f_4|g_3\rangle_2 \\
&\quad + \langle \mu f_2 + (T_{v_1} + iT_{v_2})f_3 - T_{v_3}f_4|g_4\rangle_2 = \langle A_D f|g\rangle \ ,
\end{aligned}
$$

for all $f, g \in (D(T_{v_1}) \cap D(T_{v_2}) \cap D(T_{v_3}))^4$, where $\langle\,|\,\rangle$ denotes the scalar product for $(L^2_{\mathbb{C}}(\mathbb{R}^3, \varphi_a))^2$ and $\langle\,|\,\rangle_2$ denotes the scalar product for $L^2_{\mathbb{C}}(\mathbb{R}^3, \varphi_a)$. In addition, we note that for every $\lambda \in \mathbb{C}\backslash((-\infty, -\mu] \cup [\mu, \infty))$ by

$$
B_{D\lambda} f := \frac{1}{|\,|^2 + \mu^2 - \lambda^2} \begin{pmatrix} \lambda - v_3 & -v_1 + i\,v_2 & \mu & 0 \\ -v_1 - i\,v_2 & \lambda + v_3 & 0 & \mu \\ \mu & 0 & \lambda + v_3 & v_1 - i\,v_2 \\ 0 & \mu & v_1 + i\,v_2 & \lambda - v_3 \end{pmatrix} \cdot \begin{pmatrix} f_1 \\ f_2 \\ f_3 \\ f_4 \end{pmatrix} ,
$$

for every $f \in (L^2_{\mathbb{C}}(\mathbb{R}^3, \varphi_a))^4$, there is defined a bounded linear operator on $(L^2_{\mathbb{C}}(\mathbb{R}^3, \varphi_a))^4$ such that $B_{D\lambda} f \in (D(T_{v_1}) \cap D(T_{v_2}) \cap D(T_{v_3}))^4$ and

$$
(A_D - \lambda) B_{D\lambda} f = f \ ,
$$

for every $f \in (L^2_{\mathbb{C}}(\mathbb{R}^3, \varphi_a))^4$. Hence, see, e.g., Corollary 12.4.10 in the Appendix of [7], A_D is self-adjoint, and it follows that

$$
(A_D - \lambda)^{-1} = B_{D\lambda}
$$

and hence that the spectrum $\sigma(A_D)$ of A_D satisfies

$$
\sigma(A_D) \subset (-\infty, -\mu] \cup [\mu, \infty) \ .
$$

As a consequence, by

$$
A_D^2 f := A_D A_D f \ ,
$$

for every $f \in D(A_D^2) = \{f \in D(A_D) : A_D f \in D(A_D)\}$, where $D(A_D)$ denotes the domain of A_D, there is defined a densely, linear and positive self-adjoint operator A_D^2 in $(L^2_{\mathbb{C}}(\mathbb{R}^3, \varphi_a))^2$. In particular, we have that

$$A_D^2 f = \begin{pmatrix} T_{|\ |^2+\mu^2} f_1 \\ T_{|\ |^2+\mu^2} f_2 \\ T_{|\ |^2+\mu^2} f_3 \\ T_{|\ |^2+\mu^2} f_4 \end{pmatrix} , \tag{1.30}$$

for every $f \in D(A_D^2)$, where $T_{|\ |^2+\mu^2}$ denotes the maximal multiplication operator in $L^2_{\mathbb{C}}(\mathbb{R}^3, \varphi_a)$ with $|\ |^2 + \mu^2$. We note that if $f, g \in D(A_D^2)$, then the paths $u, v : \mathbb{R} \to (L^2_{\mathbb{C}}(\mathbb{R}^3, \varphi_a))^2$, for every $t \in \mathbb{R}$ defined by

$$u(t) := e^{itA_D} f \ , \ v(t) := e^{-itA_D} g \ ,$$

assume their values in $D(A_D^2)$ and are twice differentiable, with derivatives

$$u''(t) = -A_D^2 u(t) \ , \ v''(t) = -A_D^2 v(t) \ , \tag{1.31}$$

for every $t \in \mathbb{R}$, i.e., essentially, the components of u and v satisfy a wave equation, describing the propagation of a massive field in Minkowski space. In this connection, we remind that $L^2_{\mathbb{C}}(\mathbb{R}^3, \varphi_a)$ might be interpreted as momentum space. The observation (1.31) can be used to formulate a well-posed initial value problem for the wave equation

$$u''(t) = -A_D u(t) \ ,$$

$t \in \mathbb{R}$, for general densely-defined, linear and semi-bounded self-adjoint operators A_D in Hilbert spaces. For details, we refer to [4], Theorem 2.2.1 and Corollary 2.2.2.

Since A_D leaves $(C_0(\mathbb{R}^3, \mathbb{C}))^4$ invariant, we have that $(C_0(\mathbb{R}^3, \mathbb{C}))^4$ is part of the C^∞-vectors of A_D,

$$(C_0(\mathbb{R}^3, \mathbb{C}))^4 \subset C^\infty(A_D) \ .$$

In addition, for $f \in (C_0(\mathbb{R}^3, \mathbb{C}))^4$, we have that

$$\begin{aligned}
&\|A_D f\|^2 \\
&= \| -v_3 f_1 - (v_1 - iv_2) f_2 + \mu f_3 \|_2^2 + \| -(v_1 + iv_2) f_1 + v_3 f_2 + \mu f_4 \|_2^2 \\
&\quad + \|\mu f_1 + v_3 f_3 + (v_1 - iv_2) f_4\|_2^2 + \|\mu f_2 + (v_1 + iv_2) f_3 - v_3 f_4\|_2^2 \\
&\leqslant 4\,[\, \|v_3 f_1\|_2^2 + \|(v_1 - iv_2) f_2\|_2^2 + \mu^2 \|f_3\|_2^2] \\
&\quad + 4\,[\, \|(v_1 + iv_2) f_1\|_2^2 + \|v_3 f_2\|_2^2 + \mu^2 \|f_4\|_2^2 \,] \\
&\quad + 4\,[\, \mu^2 \|f_1\|_2^2 + \|v_3 f_3\|_2^2 + \|(v_1 - iv_2) f_4\|_2^2 \,]
\end{aligned}$$

$$+ 4\,[\,\mu^2 \|\, f_2 \|_2^2 + \|(v_1 + i v_2)\, f_3 \|_2^2 + \| v_3 f_4 \|_2^2\,]$$
$$\leqslant 4\,(\mu^2 + 2R^2)\, \| f \|^2$$

and hence that

$$\| A_D f \| \leqslant 2\,(\mu^2 + 2R^2)^{1/2}\, \| f \| \,,$$

where $R > 0$ is such that $\text{supp}(f_1), \text{supp}(f_2), \text{supp}(f_3), \text{supp}(f_4) \subset U_R(0)$. Since A_D leaves the support of the components of f unchanged, from the latter it follows that

$$\| A_D^k f \| \leqslant [2\,(\mu^2 + 2R^2)^{1/2}]^k\, \| f \| \,,$$

for every $k \in \mathbb{N}$. Hence, for every $t > 0$, the sequence

$$\left(\frac{t^k}{k!}\, A_D^k f \right)_{k \in \mathbb{N}}$$

is absolutely summable, and therefore f is an analytic vector for A_D. As a consequence, for every $t \in \mathbb{R}$, we have that

$$\exp(itA_D) f = \sum_{k=0}^{\infty} \frac{(it)^k}{k!}\, A_D^k f = \sum_{k=0}^{\infty} \frac{(it)^{2k}}{(2k)!}\, A_D^{2k} f + \sum_{k=0}^{\infty} \frac{(it)^{2k+1}}{(2k+1)!}\, A_D^{2k+1} f$$

$$= \sum_{k=0}^{\infty} (-1)^k \frac{t^{2k}}{(2k)!}\, A_D^{2k} f + i \sum_{k=0}^{\infty} (-1)^k \frac{t^{2k+1}}{(2k+1)!}\, A_D^{2k} A_D f$$

$$= \sum_{k=0}^{\infty} (-1)^k \frac{t^{2k}}{(2k)!} \begin{pmatrix} [\,|\;|^2 + \mu^2]^k f_1 \\ [\,|\;|^2 + \mu^2]^k f_2 \\ [\,|\;|^2 + \mu^2]^k f_3 \\ [\,|\;|^2 + \mu^2]^k f_4 \end{pmatrix}$$

$$+ i \sum_{k=0}^{\infty} (-1)^k \frac{t^{2k+1}}{(2k+1)!} \begin{pmatrix} [\,|\;|^2 + \mu^2]^k (A_D f)_1 \\ [\,|\;|^2 + \mu^2]^k (A_D f)_2 \\ [\,|\;|^2 + \mu^2]^k (A_D f)_3 \\ [\,|\;|^2 + \mu^2]^k (A_D f)_4 \end{pmatrix}$$

$$= \begin{pmatrix} \cos(t\,[\,|\;|^2 + \mu^2]^{1/2})\, f_1 \\ \cos(t\,[\,|\;|^2 + \mu^2]^{1/2})\, f_2 \\ \cos(t\,[\,|\;|^2 + \mu^2]^{1/2})\, f_3 \\ \cos(t\,[\,|\;|^2 + \mu^2]^{1/2})\, f_4 \end{pmatrix} + i \begin{pmatrix} \frac{\sin(t\,[\,|\;|^2+\mu^2]^{1/2})}{[\,|\;|^2+\mu^2]^{1/2}}\, (A_D f)_1 \\ \frac{\sin(t\,[\,|\;|^2+\mu^2]^{1/2})}{[\,|\;|^2+\mu^2]^{1/2}}\, (A_D f)_2 \\ \frac{\sin(t\,[\,|\;|^2+\mu^2]^{1/2})}{[\,|\;|^2+\mu^2]^{1/2}}\, (A_D f)_3 \\ \frac{\sin(t\,[\,|\;|^2+\mu^2]^{1/2})}{[\,|\;|^2+\mu^2]^{1/2}}\, (A_D f)_4 \end{pmatrix} ,$$

where we used (1.26). We note that this implies that $\exp(itA_D) f \in (C_0(\mathbb{R}^3, \mathbb{C}))^4$ and therefore that $(C_0(\mathbb{R}^3, \mathbb{C}))^4$ is a core for A_D. Hence, we arrive at the following result.

For every $t \in \mathbb{R}$ and $f \in D(A_D)$, we have that

$$\exp(itA_D)f = \begin{pmatrix} \cos(t\,[|\ |^2 + \mu^2]^{1/2})\,f_1 \\ \cos(t\,[|\ |^2 + \mu^2]^{1/2})\,f_2 \\ \cos(t\,[|\ |^2 + \mu^2]^{1/2})\,f_3 \\ \cos(t\,[|\ |^2 + \mu^2]^{1/2})\,f_4 \end{pmatrix} + i \begin{pmatrix} \frac{\sin(t\,[|\ |^2+\mu^2]^{1/2})}{[|\ |^2+\mu^2]^{1/2}}\,(A_D f)_1 \\ \frac{\sin(t\,[|\ |^2+\mu^2]^{1/2})}{[|\ |^2+\mu^2]^{1/2}}\,(A_D f)_2 \\ \frac{\sin(t\,[|\ |^2+\mu^2]^{1/2})}{[|\ |^2+\mu^2]^{1/2}}\,(A_D f)_3 \\ \frac{\sin(t\,[|\ |^2+\mu^2]^{1/2})}{[|\ |^2+\mu^2]^{1/2}}\,(A_D f)_4 \end{pmatrix} .$$

In the following, we are going to show that the spectrum $\sigma(A_D)$ of A_D is given by all real numbers. For this purpose, for $\nu \in \mathbb{N}^*$, we define the auxiliary function $k_\nu \in C_0(\mathbb{R}^3, \mathbb{C})$ by

$$k_\nu(v) := \begin{cases} 0 & \text{if } |v| \geqslant 1/\nu, \\ 1 - \nu^2|v|^2 & \text{if } |v| < 1/\nu. \end{cases}$$

for every $v \in \mathbb{R}^3$. Then $k_\nu(0) = 1$, $0 \leqslant k_\nu \leqslant 1$, $\text{supp}(k_\nu) \subset B_{1/\nu}(0)$. Hence for $\mathfrak{v}_0 \in \mathbb{R}^3 \setminus \{0\}$, we have that

$$f_{\mathfrak{v}_0,\nu} := k_\nu(\text{id}_{\mathbb{R}^3} - \mathfrak{v}_0) \in C_0(\mathbb{R}^3, \mathbb{C}) \ ,$$

$f_{\mathfrak{v}_0,\nu}(\mathfrak{v}_0) = 1$, $0 \leqslant f_{\mathfrak{v}_0,\nu} \leqslant 1$, $\text{supp}(f_{\mathfrak{v}_0,\nu}) \subset B_{1/\nu}(\mathfrak{v}_0)$. In the following, let $\lambda \in ((-\infty, -\mu) \cup (\mu, \infty))$ and $\mathfrak{v}_0 \in \mathbb{R}^3 \setminus \{0\}$ such that

$$\begin{cases} (|\mathfrak{v}_0|^2 + \mu^2)^{1/2} = \lambda & \text{if } \lambda > 0 \\ (|\mathfrak{v}_0|^2 + \mu^2)^{1/2} = -\lambda & \text{if } \lambda < 0 \end{cases} .$$

If $\lambda > 0$, we have that

$$\begin{aligned} &(A_D - \lambda) \left[f_{\mathfrak{v}_0,\nu} \cdot \begin{pmatrix} v_3 - (|\ |^2 + \mu^2)^{1/2} \\ v_1 + i\, v_2 \\ -\mu \\ 0 \end{pmatrix} \right] \\ &= f_{\mathfrak{v}_0,\nu} \cdot [(|\ |^2 + \mu^2)^{1/2} - \lambda] \cdot \begin{pmatrix} v_3 - (|\ |^2 + \mu^2)^{1/2} \\ v_1 + i\, v_2 \\ -\mu \\ 0 \end{pmatrix} \\ &= f_{\mathfrak{v}_0,\nu} \cdot [(|\ |^2 + \mu^2)^{1/2} - (|\mathfrak{v}_0|^2 + \mu^2)^{1/2}] \cdot \begin{pmatrix} v_3 - (|\ |^2 + \mu^2)^{1/2} \\ v_1 + i\, v_2 \\ -\mu \\ 0 \end{pmatrix} , \end{aligned}$$

whereas if $\lambda < 0$, we have that

$$(A_D - \lambda)\left[f_{\mathfrak{v}_0,\nu} \cdot \begin{pmatrix} v_3 + (|\;|^2 + \mu^2)^{1/2} \\ v_1 + i\, v_2 \\ -\mu \\ 0 \end{pmatrix} \right]$$

$$= - f_{\mathfrak{v}_0,\nu} \cdot [\,(|\;|^2 + \mu^2)^{1/2} + \lambda\,] \cdot \begin{pmatrix} v_3 + (|\;|^2 + \mu^2)^{1/2} \\ v_1 + i\, v_2 \\ -\mu \\ 0 \end{pmatrix}$$

$$= - f_{\mathfrak{v}_0,\nu} \cdot [\,(|\;|^2 + \mu^2)^{1/2} - (|\mathfrak{v}_0|^2 + \mu^2)^{1/2}\,] \cdot \begin{pmatrix} v_3 + (|\;|^2 + \mu^2)^{1/2} \\ v_1 + i\, v_2 \\ -\mu \\ 0 \end{pmatrix} .$$

Since for $v \in B_{1/\nu}(\mathfrak{v}_0)$, we have that

$$|\,|v| - |\mathfrak{v}_0|\,| \leqslant |v - \mathfrak{v}_0| \leqslant \frac{1}{\nu} ,$$

$$|(\,|v|^2 + \mu^2)^{1/2} - (|\mathfrak{v}_0|^2 + \mu^2)^{1/2}| = \frac{|\,|v|^2 - |\mathfrak{v}_0|^2\,|}{(\,|v|^2 + \mu^2)^{1/2} + (|\mathfrak{v}_0|^2 + \mu^2)^{1/2}}$$

$$= \frac{|\,|v| - |\mathfrak{v}_0|\,| \cdot (\,|v| + |\mathfrak{v}_0|\,)}{(\,|v|^2 + \mu^2)^{1/2} + (|\mathfrak{v}_0|^2 + \mu^2)^{1/2}} \leqslant \frac{|v - \mathfrak{v}_0| \cdot (\,|v| + |\mathfrak{v}_0|\,)}{(\,|v|^2 + \mu^2)^{1/2} + (\,|\mathfrak{v}_0|^2 + \mu^2)^{1/2}}$$

$$\leqslant \frac{1}{\nu} \frac{|v| + |\mathfrak{v}_0|}{(\,|v|^2 + \mu^2)^{1/2} + (\,|\mathfrak{v}_0|^2 + \mu^2)^{1/2}} \leqslant \frac{1}{\nu} ,$$

we infer that

$$\left\| f_{\mathfrak{v}_0,\nu} \cdot [\,(|\;|^2 + \mu^2)^{1/2} - (|\mathfrak{v}_0|^2 + \mu^2)^{1/2}\,] \cdot \begin{pmatrix} v_3 \mp (|\;|^2 + \mu^2)^{1/2} \\ v_1 + i\, v_2 \\ -\mu \\ 0 \end{pmatrix} \right\|^2$$

$$= \int_{\mathbb{R}^3} (v_a^0)^{-1}\, |\,(|\;|^2 + \mu^2)^{1/2} - (|\mathfrak{v}_0|^2 + \mu^2)^{1/2}\,|^2$$

$$\left[(v_3 \mp (|\;|^2 + \mu^2)^{1/2})^2 + v_1^2 + v_2^2 + \mu^2 \right] |f_{\mathfrak{v}_0,\nu}|^2 dv^3$$

$$\leqslant \frac{1}{\nu^2} \int_{B_{1/\nu}(\mathfrak{v}_0)} (v_a^0)^{-1} \left[(v_3 \mp (|\;|^2 + \mu^2)^{1/2})^2 + v_1^2 + v_2^2 + \mu^2 \right] |f_{\mathfrak{v}_0,\nu}|^2 dv^3$$

$$= \frac{1}{\nu^2} \left\| f_{\mathfrak{v}_0,\nu} \cdot \begin{pmatrix} v_3 \mp (|\;|^2 + \mu^2)^{1/2} \\ v_1 + i\, v_2 \\ -\mu \\ 0 \end{pmatrix} \right\|^2$$

and hence that

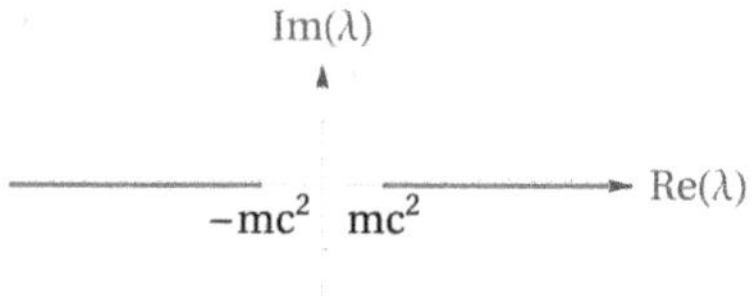

Fig. 1.10 Depiction of the spectrum (in red) of H_D

$$\lim_{\nu\to\infty} \frac{\left\| (A_D - \lambda)\left[f_{\mathfrak{v}_0,\nu} \cdot \begin{pmatrix} v_3 \mp (|\ \ |^2 + \mu^2)^{1/2} \\ v_1 + i\, v_2 \\ -\mu \\ 0 \end{pmatrix} \right] \right\|}{\left\| f_{\mathfrak{v}_0,\nu} \cdot \begin{pmatrix} v_3 \mp (|\ \ |^2 + \mu^2)^{1/2} \\ v_1 + i\, v_2 \\ -\mu \\ 0 \end{pmatrix} \right\|} = 0\ .$$

Therefore, see, e.g., Theorem 12.5.3 in the Appendix of [7], it follows that $\lambda \in \sigma(A_D)$. As a consequence, $(-\infty, -\mu) \cup (\mu, \infty) \subset \sigma(A_D)$ and, since $\sigma(A_D)$ is a closed subset of $\mathbb{R}$, that $\sigma(A_D) = (-\infty, -\mu] \cup [\mu, \infty)$.

The spectrum of A_D is given by $\sigma(A_D) = (-\infty, -\mu] \cup [\mu, \infty)$.

Therefore, the spectrum of the Hamilton operator H_D is not bounded from below. As a consequence, the Dirac equation is not suitable for the definition of a one-particle theory (Fig. 1.10).

2 A Representation Associated with the Electromagnetic Field

2.1 Transformation Properties of the Electromagnetic Field

In the following, we consider the transformation properties of a skew-symmetric tensor field F of type $(2, 0)$ on a differentiable manifold.[1] For this purpose, we assume that M is an n-dimensional differentiable (C^{∞}-) manifold, where $n \in \mathbb{N}^*$, that $f : M \to M$ is a diffeomorphism and that $p \in M$. In addition, we assume that u and $\tilde{u}$ belong to the atlas of M, with domains $D(u)$ and $D(\tilde{u})$ containing p and $f^{-1}(p)$, respectively. Then,

$$F|_{D(u)} = \sum_{k,l=1}^{n} F_{kl}\, du_k \otimes du_l \ , \quad F|_{D(\tilde{u})} = \sum_{k,l=1}^{n} \tilde{F}_{kl}\, d\tilde{u}_k \otimes d\tilde{u}_l \ ,$$

where $(F_{kl})_{k,l \in \{1,\dots,n\}}$ and $(\tilde{F}_{kl})_{k,l \in \{1,\dots,n\}}$ are families of functions defined on $D(u)$ and $D(\tilde{u})$, respectively, such that $F_{lk} = -F_{kl}$ and $\tilde{F}_{lk} = -\tilde{F}_{kl}$, for $k, l \in \{1, \dots, n\}$.[2] Motivated by the push forward of vector fields, defined in Sect. 4.5 of part I, we are going to define the push forward $f_* F$ of F such that

$$(f_* F)(f_{*f^{-1}(p)}\bar{v}, f_{*f^{-1}(p)}\bar{w}) = F(\bar{v}, \bar{w}) \ ,$$

for every $\bar{v}, \bar{w} \in T_{f^{-1}(p)}M$. Since,

---

[1] In physics, such tensor fields are used to describe electromagnetic fields within the framework of special and general relativity.

[2] If F is an electromagnetic field tensor, then $F_{01} = E_1$, $F_{02} = E_2$, $F_{03} = E_3$, $F_{32} = B_1$, $F_{13} = B_2$ and $F_{21} = B_3$, where E and B are the electric and the magnetic field, respectively. All other components of F vanish due to the fact that F is skew-symmetric. In this case, all components have the dimension $m^{1/2} l^{-1/2} t^{-1}$, where it is assumed that the coordinate projections $u_1, \dots, u_n$ have the dimension of a length.

H. R. Beyer, *Quantum Spin and Representations of the Poincaré Group, Part II*,
Synthesis Lectures on Engineering, Science, and Technology,
https://doi.org/10.1007/978-3-031-95823-6_2

$$\mathrm{id}_{T_pM} = (f \circ f^{-1})_{*p} = f_{*f^{-1}(p)} \circ (f^{-1})_{*p} ,$$
$$\mathrm{id}_{T_{f^{-1}(p)}M} = (f^{-1} \circ f)_{*f^{-1}(p)} = (f^{-1})_{*p} \circ f_{*f^{-1}(p)} ,$$

and hence

$$f_{*f^{-1}(p)} = [(f^{-1})_{*p}]^{-1} ,$$

this implies that

$$\begin{aligned}(f_*F)(v, w) &= (f_*F)(f_{*f^{-1}(p)}(f^{-1})_{*p}v, f_{*f^{-1}(p)}(f^{-1})_{*p}w) \\ &= F((f^{-1})_{*p}v, (f^{-1})_{*p}w) ,\end{aligned}$$

for all $v, w \in T_pM$. Hence, in the following, we define

$$(f_*F)(v, w) := F((f^{-1})_{*p}v, (f^{-1})_{*p}w) ,$$

for all $v, w \in T_pM$. Then, indeed,

$$\begin{aligned}(f_*F)(f_{*f^{-1}(p)}\bar{v}, f_{*f^{-1}(p)}\bar{w}) &= F((f^{-1})_{*p}f_{*f^{-1}(p)}\bar{v}, (f^{-1})_{*p}f_{*f^{-1}(p)}\bar{w}) \\ &= F(\bar{v}, \bar{w}) ,\end{aligned}$$

for all $\bar{v}, \bar{w} \in T_{f^{-1}(p)}M$. Further, if $v, w \in T_pM$, then

$$(f^{-1})_{*p}v = \sum_{k=1}^{n} \left[\sum_{l=1}^{n} \frac{\partial\,(\tilde{u}_k \circ f^{-1} \circ u^{-1})}{\partial x_l}(u(p))\, v_l \right] \frac{\partial}{\partial \tilde{u}_k}\bigg|_{f^{-1}(p)} ,$$

where $v_1, \ldots, v_n$ are the coefficients in the expansion of v, in terms of the basis $\frac{\partial}{\partial u_1}|_p, \ldots, \frac{\partial}{\partial u_n}|_p$ of T_pM, and hence,

$$\begin{aligned}(f_*F)(v, w) &:= F((f^{-1})_{*p}v, (f^{-1})_{*p}w) \\ &= \sum_{k,l,\bar{k},\bar{l}=1}^{n} \frac{\partial\,(\tilde{u}_k \circ f^{-1} \circ u^{-1})}{\partial x_l}(u(p))\, \frac{\partial\,(\tilde{u}_{\bar{k}} \circ f^{-1} \circ u^{-1})}{\partial x_{\bar{l}}}(u(p))\, v_l\, w_{\bar{l}} \\ &\qquad F\left(\frac{\partial}{\partial \tilde{u}_k}\bigg|_{f^{-1}(p)}, \frac{\partial}{\partial \tilde{u}_{\bar{k}}}\bigg|_{f^{-1}(p)} \right) \\ &= \sum_{k,l,\bar{k},\bar{l}=1}^{n} \frac{\partial\,(\tilde{u}_k \circ f^{-1} \circ u^{-1})}{\partial x_l}(u(p))\, \frac{\partial\,(\tilde{u}_{\bar{k}} \circ f^{-1} \circ u^{-1})}{\partial x_{\bar{l}}}(u(p)) \\ &\qquad \tilde{F}_{k\bar{k}}(f^{-1}(p))\, v_l\, w_{\bar{l}} ,\end{aligned}$$

where $w_1, \ldots, w_n$ are the coefficients in the expansion of w, in terms of the basis $\frac{\partial}{\partial u_1}|_p, \ldots, \frac{\partial}{\partial u_n}|_p$ of T_pM. In particular, if $M = \mathbb{R}^n$, $f : \mathbb{R}^n \to \mathbb{R}^n$ is given by

$$f(x) := \Lambda \cdot x = {}^t\left(\sum_{m=1}^{n} \Lambda_{1m} x_m, \dots, \sum_{k=1}^{n} \Lambda_{nm} x_m\right),$$

for every $x = {}^t(x_1, \dots, x_n) \in \mathbb{R}^n$, where $\Lambda \in \mathrm{GL}(n, \mathbb{R})$, $u = \tilde{u} = \mathrm{id}_{\mathbb{R}^n}$, then

$$(f_* F)(v, w) = \sum_{k,l,\bar{k},\bar{l}=1}^{n} \Lambda^{-1}_{\bar{k}l} \Lambda^{-1}_{\bar{k}\bar{l}} \, F_{k\bar{k}}(\Lambda^{-1} p) \, v_l w_{\bar{l}}$$

and hence

$$\begin{aligned}
(f^* F)_p &= \sum_{k,l,\bar{k},\bar{l}=1}^{n} \Lambda^{-1}_{\bar{k}l} \Lambda^{-1}_{\bar{k}\bar{l}} \, F_{k\bar{k}}(\Lambda^{-1} p) \, dx_l \otimes dx_{\bar{l}}\big|_p \\
&= \sum_{k,l,m,\bar{m}=1}^{n} \Lambda^{-1}_{mk} \Lambda^{-1}_{\bar{m}l} \, F_{m\bar{m}}(\Lambda^{-1} p) \, dx_k \otimes dx_l\big|_p \\
&= \sum_{k,l=1}^{n} \left[\sum_{m,\bar{m}=1}^{n} \Lambda^{-1}_{mk} \Lambda^{-1}_{\bar{m}l} \, F_{m\bar{m}}(\Lambda^{-1} p) \right] dx_k \otimes dx_l\big|_p \\
&= \sum_{k,l=1}^{n} \left[\sum_{m,\bar{m}=1}^{n} [(\Lambda^{-1})^t]_{km} \, F_{m\bar{m}}(\Lambda^{-1} p) \, \Lambda^{-1}_{\bar{m}l} \right] dx_k \otimes dx_l\big|_p \\
&= \sum_{k,l=1}^{n} \left([(\Lambda^{-1})^t] \cdot (F_{m\bar{m}}(\Lambda^{-1} p))_{m,\bar{m} \in \{1,\dots,n\}} \cdot \Lambda^{-1} \right)_{kl} dx_k \otimes dx_l\big|_p ,
\end{aligned}$$

where $x_1, \dots, x_n$ denote the coordinate projections of $\mathbb{R}^n$. Hence, the push forward

$$F_\Lambda(p) := (f^* F)_p$$

of $F(p)$ under the action of Λ is given by

$$F_\Lambda(p) = \sum_{k,l=1}^{n} \left([(\Lambda^{-1})^t] \cdot (F_{m\bar{m}}(\Lambda^{-1} p))_{m,\bar{m} \in \{1,\dots,n\}} \cdot \Lambda^{-1} \right)_{kl} dx_k \otimes dx_l\big|_p ,$$

for every $p \in \mathbb{R}^n$.

As a consequence, for every $\Lambda \in \mathrm{GL}(n, \mathbb{R})$ and $p \in \mathbb{R}^n$, the coefficicients in the expansions of the push forward $F_\Lambda(p)$, under the action of Λ, and of $F(p)$, in terms of the basis $(dx_k \otimes dx_l\big|_p)_{k,l \in \{1,\dots,n\}}$ for all bilinear forms on $T_p M$, are related by

$$(F_{\Lambda kl}(p))_{k,l \in \{1,\dots,n\}} = [(\Lambda^{-1})^t] \cdot (F_{kl}(\Lambda^{-1} p))_{k,l \in \{1,\dots,n\}} \cdot \Lambda^{-1} ,$$

where $x_1, \dots, x_n$ denote the coordinate projections of $\mathbb{R}^n$.

2.2 A Representation of $\mathrm{GL}(4, \mathbb{R})$

The transformation properties of the electromagnetic field, derived in the previous chapter, are linked to the representation

$$\mathfrak{R} : \mathrm{GL}(4, \mathbb{R}) \to \mathrm{L}(\mathrm{Skew}(4, \mathbb{C})) \ ,$$

where $\mathrm{Skew}(4, \mathbb{C})$ denotes the 6-dimensional complex vector space of skew-symmetric complex 4×4 matrices,

$$\mathrm{Skew}(4, \mathbb{C}) := \{F \in M(4, \mathbb{C}) : F^t = -F\} \ ,$$

defined by

$$\mathfrak{R}(S) := (\mathrm{Skew}(4, \mathbb{C}) \to \mathrm{Skew}(4, \mathbb{C}), F \mapsto (S^{-1})^t \cdot F \cdot S^{-1}) \ ,$$

for every $S \in \mathrm{GL}(4, \mathbb{R})$. For $S \in \mathrm{GL}(4, \mathbb{R})$, $\mathfrak{R}(S)$ is well-defined, since

$$\left((S^{-1})^t \cdot F \cdot S^{-1}\right)^t = (S^{-1})^t \cdot F^t \cdot (S^{-1})^{tt} = -(S^{-1})^t \cdot F \cdot S^{-1} \ ,$$

for every $F \in \mathrm{Skew}(4, \mathbb{C})$, and, as a consequence of the linearity of matrix multiplication, linear. Further, for $S_1, S_2 \in \mathrm{GL}(4, \mathbb{R})$, we have that

$$\begin{aligned}
&\mathfrak{R}(S_1 \cdot S_2)F = ((S_1 S_2)^{-1})^t \cdot F \cdot (S_1 \cdot S_2)^{-1} \\
&= (S_1^{-1})^t \cdot (S_2^{-1})^t \cdot F \cdot S_2^{-1} \cdot S_1^{-1} = (S_1^{-1})^t \cdot (\mathfrak{R}(S_2)F) \cdot S_1^{-1} \\
&= \mathfrak{R}(S_1)\mathfrak{R}(S_2)F = (\mathfrak{R}(S_1) \circ \mathfrak{R}(S_2))F
\end{aligned}$$

and that

$$\mathfrak{R}(E)F = (E^{-1})^t \cdot F \cdot E^{-1} = F \ ,$$

for every $F \in \mathrm{Skew}(4, \mathbb{C})$, and hence that

$$\mathfrak{R}(S_1 \cdot S_2) = \mathfrak{R}(S_1) \circ \mathfrak{R}(S_2) \ , \ \mathfrak{R}(E) = \mathrm{id}_{\mathrm{Skew}(4,\mathbb{C})} \ .$$

Therefore, $\mathfrak{R}$ is a representation of $\mathrm{GL}(4, \mathbb{R})$. We note that for every $F \in \mathrm{Skew}(4, \mathbb{C})$, we have that

$$F = \begin{pmatrix} 0 & F_{01} & F_{02} & F_{03} \\ -F_{01} & 0 & -F_{21} & F_{13} \\ -F_{02} & F_{21} & 0 & -F_{32} \\ -F_{03} & -F_{13} & F_{32} & 0 \end{pmatrix} \ .$$

Exercise 2.1 Show that by

$$\begin{aligned}\langle F, G\rangle &:= \frac{1}{2}.\text{Tr}(F^*G) \\ &= F_{01}^*G_{01} + F_{02}^*G_{02} + F_{03}^*G_{03} + F_{32}^*G_{32} + F_{13}^*G_{13} + F_{21}^*G_{21} \ ,\end{aligned} \tag{2.1}$$

for every

$$F = \begin{pmatrix} 0 & F_{01} & F_{02} & F_{03} \\ -F_{01} & 0 & -F_{21} & F_{13} \\ -F_{02} & F_{21} & 0 & -F_{32} \\ -F_{03} & -F_{13} & F_{32} & 0 \end{pmatrix} , \ G = \begin{pmatrix} 0 & G_{01} & G_{02} & G_{03} \\ -G_{01} & 0 & -G_{21} & G_{13} \\ -G_{02} & G_{21} & 0 & -G_{32} \\ -G_{03} & -G_{13} & G_{32} & 0 \end{pmatrix}$$
$$\in \text{Skew}(4, \mathbb{C}) \ ,$$

there is defined a scalar product $\langle\ ,\ \rangle : \text{Skew}(4, \mathbb{C}) \times \text{Skew}(4, \mathbb{C}) \to \mathbb{C}$ in $\text{Skew}(4, \mathbb{C})$.

In the following, we consider $\text{Skew}(4, \mathbb{C})$ equipped with the norm topology that is induced by the norm $|\ |_{\text{Tr}} : \text{Skew}(4, \mathbb{C}) \to \mathbb{R}$, defined by $|F|_{\text{Tr}} := \langle F, F\rangle^{1/2}$, for every $F \in \text{Skew}(4, \mathbb{C})$.

For $S \in \text{GL}(4, \mathbb{R})$, $F_1, F_2 \in \text{Skew}(4, \mathbb{C})$, we have that

$$\begin{aligned}&\langle \mathfrak{R}(S)F_1, \mathfrak{R}(S)F_2\rangle = \langle (S^{-1})^t \cdot F_1 \cdot S^{-1}, (S^{-1})^t \cdot F_2 \cdot S^{-1}\rangle \\ &= \frac{1}{2}.\text{Tr}(((S^{-1})^t \cdot F_1 \cdot S^{-1})^* \cdot (S^{-1})^t \cdot F_2 \cdot S^{-1}) \\ &= \frac{1}{2}.\text{Tr}((S^{-1})^t \cdot F_1^* \cdot S^{-1} \cdot (S^{-1})^t \cdot F_2 \cdot S^{-1}) \\ &= \frac{1}{2}.\text{Tr}((F_1^* \cdot S^{-1} \cdot (S^{-1})^t \cdot F_2 \cdot S^{-1} \cdot (S^{-1})^t)\end{aligned}$$

and hence that $\mathfrak{R}(S)$ preserves the scalar product and hence is unitary if

$$S^{-1} \cdot (S^{-1})^t = E \ .$$

Since

$$S^{-1} \cdot (S^{-1})^t = E \Leftrightarrow S = (S^{-1})^t \Leftrightarrow S^t = S^{-1} \Leftrightarrow S \in \text{O}(4) \ ,$$

it follows that if $S \in \mathrm{O}(4)$, then $\mathfrak{R}(S)$ is unitary.

Exercise 2.2 Show directly that the maximum norm $\| \ \|_\infty$ and $| \ |_{\mathrm{Tr}}$ are equivalent.

Exercise 2.3 Show that $\mathfrak{R}$ is continuous.

Exercise 2.4 Show that by[3]

$$\mathfrak{I}(\psi) := \begin{pmatrix} 0 & \psi_1 & \psi_2 & \psi_3 \\ -\psi_1 & 0 & -\psi_6 & \psi_5 \\ -\psi_2 & \psi_6 & 0 & -\psi_4 \\ -\psi_3 & -\psi_5 & \psi_4 & 0 \end{pmatrix},$$

for every $\psi \in \mathbb{C}^6$, there is defined an a Hilbert space isomorphism $\mathfrak{I} : \mathbb{C}^6 \to \mathrm{Skew}(4, \mathbb{C})$, where $\mathbb{C}^6$ is equipped with the canonical scalar product, with inverse

$$\mathfrak{I}^{-1}(F) := \begin{pmatrix} F_{01} \\ F_{02} \\ F_{03} \\ F_{32} \\ F_{13} \\ F_{21} \end{pmatrix},$$

for every $F \in \mathrm{Skew}(4, \mathbb{C})$.

From the result of Exercise 2.4 and the fact that $\mathfrak{R}$ is a representation of $\mathrm{GL}(4, \mathbb{R})$, it follows that by

$$\mathfrak{R}'(S) := \mathfrak{I}^{-1} \circ \mathfrak{R}(S) \circ \mathfrak{I} ,$$

for every $S \in \mathrm{GL}(4, \mathbb{C})$, there is defined a representation

$$\mathfrak{R}' : \mathrm{GL}(4, \mathbb{R}) \to \mathrm{L}(\mathbb{C}^6) ,$$

of $\mathrm{GL}(4, \mathbb{R})$. After some calculation, it follows for $\Lambda \in \mathcal{L}$ that

[3] If $F(p)$ is an electromagnetic field tensor at a point $p \in \mathbb{R}^4$, then $F_{01}(p) = E_1(p)$, $F_{02}(p) = E_2(p)$, $F_{03}(p) = E_3(p)$, $F_{32}(p) = B_1(p)$, $F_{13}(p) = B_2(p)$ and $F_{21}(p) = B_3(p)$, where $E(p)$ and $B(p)$ are the electric and the magnetic field at p, respectively. All other components of $F(p)$ vanish due to the fact that $F(p)$ is skew-symmetric. Then, $\mathfrak{I}(E_1(p), E_2(p), E_3(p), B_1(p), B_2(p), B_3(p)) = (F_{kl}(p))_{k,l \in \{0,\ldots,3\}} \in \mathrm{Skew}(4, \mathbb{C})$. This fact motivates the definition of the isomorphism $\mathfrak{I}$.

$$
\begin{aligned}
&(\mathfrak{R}'(\Lambda)\psi)_1 = \\
&(\Lambda_{00}\Lambda_{11} - \Lambda_{01}\Lambda_{10})\,\psi_1 + (\Lambda_{00}\Lambda_{12} - \Lambda_{02}\Lambda_{10})\,\psi_2 + (\Lambda_{00}\Lambda_{13} - \Lambda_{03}\Lambda_{10})\,\psi_3 \\
&+ (\Lambda_{02}\Lambda_{13} - \Lambda_{03}\Lambda_{12})\,\psi_4 + (\Lambda_{03}\Lambda_{11} - \Lambda_{01}\Lambda_{13})\,\psi_5 + (\Lambda_{01}\Lambda_{12} - \Lambda_{02}\Lambda_{11})\,\psi_6\,, \\
&(\mathfrak{R}'(\Lambda)\psi)_2 = \\
&(\Lambda_{00}\Lambda_{21} - \Lambda_{01}\Lambda_{20})\,\psi_1 + (\Lambda_{00}\Lambda_{22} - \Lambda_{02}\Lambda_{20})\,\psi_2 + (\Lambda_{00}\Lambda_{23} - \Lambda_{03}\Lambda_{20})\,\psi_3 \\
&+ (\Lambda_{02}\Lambda_{23} - \Lambda_{03}\Lambda_{22})\,\psi_4 + (\Lambda_{03}\Lambda_{21} - \Lambda_{01}\Lambda_{23})\,\psi_5 + (\Lambda_{01}\Lambda_{22} - \Lambda_{02}\Lambda_{21})\,\psi_6\,, \\
&(\mathfrak{R}'(\Lambda)\psi)_3 = \\
&(\Lambda_{00}\Lambda_{31} - \Lambda_{01}\Lambda_{30})\,\psi_1 + (\Lambda_{00}\Lambda_{32} - \Lambda_{02}\Lambda_{30})\,\psi_2 + (\Lambda_{00}\Lambda_{33} - \Lambda_{03}\Lambda_{30})\,\psi_3 \\
&+ (\Lambda_{02}\Lambda_{33} - \Lambda_{03}\Lambda_{32})\,\psi_4 + (\Lambda_{03}\Lambda_{31} - \Lambda_{01}\Lambda_{33})\,\psi_5 + (\Lambda_{01}\Lambda_{32} - \Lambda_{02}\Lambda_{31})\,\psi_6\,, \\
&(\mathfrak{R}'(\Lambda)\psi)_4 = \\
&(\Lambda_{20}\Lambda_{31} - \Lambda_{21}\Lambda_{30})\,\psi_1 + (\Lambda_{20}\Lambda_{32} - \Lambda_{22}\Lambda_{30})\,\psi_2 + (\Lambda_{20}\Lambda_{33} - \Lambda_{23}\Lambda_{30})\,\psi_3 \\
&+ (\Lambda_{22}\Lambda_{33} - \Lambda_{23}\Lambda_{32})\,\psi_4 + (\Lambda_{23}\Lambda_{31} - \Lambda_{21}\Lambda_{33})\,\psi_5 + (\Lambda_{21}\Lambda_{32} - \Lambda_{22}\Lambda_{31})\,\psi_6\,, \\
&(\mathfrak{R}'(\Lambda)\psi)_5 = \\
&(\Lambda_{11}\Lambda_{30} - \Lambda_{10}\Lambda_{31})\,\psi_1 + (\Lambda_{12}\Lambda_{30} - \Lambda_{10}\Lambda_{32})\,\psi_2 + (\Lambda_{13}\Lambda_{30} - \Lambda_{10}\Lambda_{33})\,\psi_3 \\
&+ (\Lambda_{13}\Lambda_{32} - \Lambda_{12}\Lambda_{33})\,\psi_4 + (\Lambda_{11}\Lambda_{33} - \Lambda_{13}\Lambda_{31})\,\psi_5 + (\Lambda_{12}\Lambda_{31} - \Lambda_{11}\Lambda_{32})\,\psi_6\,, \\
&(\mathfrak{R}'(\Lambda)\psi)_6 = \\
&(\Lambda_{10}\Lambda_{21} - \Lambda_{11}\Lambda_{20})\,\psi_1 + (\Lambda_{10}\Lambda_{22} - \Lambda_{12}\Lambda_{20})\,\psi_2 + (\Lambda_{10}\Lambda_{23} - \Lambda_{13}\Lambda_{20})\,\psi_3 \\
&+ (\Lambda_{12}\Lambda_{23} - \Lambda_{13}\Lambda_{22})\,\psi_4 + (\Lambda_{13}\Lambda_{21} - \Lambda_{11}\Lambda_{23})\,\psi_5 + (\Lambda_{11}\Lambda_{22} - \Lambda_{12}\Lambda_{21})\,\psi_6\,.
\end{aligned}
$$

and hence that the representation matrix $M_{\mathfrak{R}'(\Lambda)}$ of $\mathfrak{R}'(\Lambda)$, with respect to the canonical basis of $\mathbb{C}^6$, is given by

$$
M_{\mathfrak{R}'(\Lambda)} = \\
\begin{pmatrix}
\Lambda_{00}\Lambda_{11} - \Lambda_{01}\Lambda_{10} & \Lambda_{00}\Lambda_{12} - \Lambda_{02}\Lambda_{10} & \Lambda_{00}\Lambda_{13} - \Lambda_{03}\Lambda_{10} & \Lambda_{02}\Lambda_{13} - \Lambda_{03}\Lambda_{12} & \Lambda_{03}\Lambda_{11} - \Lambda_{01}\Lambda_{13} & \Lambda_{01}\Lambda_{12} - \Lambda_{02}\Lambda_{11} \\
\Lambda_{00}\Lambda_{21} - \Lambda_{01}\Lambda_{20} & \Lambda_{00}\Lambda_{22} - \Lambda_{02}\Lambda_{20} & \Lambda_{00}\Lambda_{23} - \Lambda_{03}\Lambda_{20} & \Lambda_{02}\Lambda_{23} - \Lambda_{03}\Lambda_{22} & \Lambda_{03}\Lambda_{21} - \Lambda_{01}\Lambda_{23} & \Lambda_{01}\Lambda_{22} - \Lambda_{02}\Lambda_{21} \\
\Lambda_{00}\Lambda_{31} - \Lambda_{01}\Lambda_{30} & \Lambda_{00}\Lambda_{32} - \Lambda_{02}\Lambda_{30} & \Lambda_{00}\Lambda_{33} - \Lambda_{03}\Lambda_{30} & \Lambda_{02}\Lambda_{33} - \Lambda_{03}\Lambda_{32} & \Lambda_{03}\Lambda_{31} - \Lambda_{01}\Lambda_{33} & \Lambda_{01}\Lambda_{32} - \Lambda_{02}\Lambda_{31} \\
\Lambda_{20}\Lambda_{31} - \Lambda_{21}\Lambda_{30} & \Lambda_{20}\Lambda_{32} - \Lambda_{22}\Lambda_{30} & \Lambda_{20}\Lambda_{33} - \Lambda_{23}\Lambda_{30} & \Lambda_{22}\Lambda_{33} - \Lambda_{23}\Lambda_{32} & \Lambda_{23}\Lambda_{31} - \Lambda_{21}\Lambda_{33} & \Lambda_{21}\Lambda_{32} - \Lambda_{22}\Lambda_{31} \\
\Lambda_{11}\Lambda_{30} - \Lambda_{10}\Lambda_{31} & \Lambda_{12}\Lambda_{30} - \Lambda_{10}\Lambda_{32} & \Lambda_{13}\Lambda_{30} - \Lambda_{10}\Lambda_{33} & \Lambda_{13}\Lambda_{32} - \Lambda_{12}\Lambda_{33} & \Lambda_{11}\Lambda_{33} - \Lambda_{13}\Lambda_{31} & \Lambda_{12}\Lambda_{31} - \Lambda_{11}\Lambda_{32} \\
\Lambda_{10}\Lambda_{21} - \Lambda_{11}\Lambda_{20} & \Lambda_{10}\Lambda_{22} - \Lambda_{12}\Lambda_{20} & \Lambda_{10}\Lambda_{23} - \Lambda_{13}\Lambda_{20} & \Lambda_{12}\Lambda_{23} - \Lambda_{13}\Lambda_{22} & \Lambda_{13}\Lambda_{21} - \Lambda_{11}\Lambda_{23} & \Lambda_{11}\Lambda_{22} - \Lambda_{12}\Lambda_{21}
\end{pmatrix}.
$$

We note that it follows that $D_{\mathfrak{M}} : \mathcal{L} \to \mathrm{GL}(6, \mathbb{R})$, defined by

$$D_{\mathfrak{M}}(\Lambda) := M_{\mathfrak{R}'(\Lambda)}\,,$$

for every $\Lambda \in \mathcal{L}$, is a representation of $\mathcal{L}$. In particular, we have that

$$\|D_{\mathfrak{M}}(\Lambda)\|_\infty \leqslant 2\,\|\Lambda\|_\infty^2\,,$$

for every $\Lambda \in \mathcal{L}$. Further, since for $\Lambda_1, \Lambda_2 \in \mathcal{L}$,

$$\Lambda_{1jk}\Lambda_{1lm} - \Lambda_{2jk}\Lambda_{2lm} = (\Lambda_{1jk} - \Lambda_{2jk})\Lambda_{1lm} + \Lambda_{2jk}(\Lambda_{1lm} - \Lambda_{2lm})$$
$$= (\Lambda_{1jk} - \Lambda_{2jk})\Lambda_{1lm} + (\Lambda_{2jk} - \Lambda_{1jk})(\Lambda_{1lm} - \Lambda_{2lm}) + \Lambda_{1jk}(\Lambda_{1lm} - \Lambda_{2lm})$$

and hence

$$\begin{aligned}
&|\Lambda_{1jk}\Lambda_{1lm} - \Lambda_{2jk}\Lambda_{2lm}| \\
&\leqslant |\Lambda_{1jk} - \Lambda_{2jk}| \cdot |\Lambda_{1lm}| + |\Lambda_{1jk} - \Lambda_{2jk}| \cdot |\Lambda_{1lm} - \Lambda_{2lm}| \\
&\quad + |\Lambda_{1jk}| \cdot |\Lambda_{1lm} - \Lambda_{2lm}| \\
&\leqslant 2\|\Lambda_1\|_\infty \cdot \|\Lambda_1 - \Lambda_2\|_\infty + \|\Lambda_1 - \Lambda_2\|_\infty^2 ,
\end{aligned}$$

for all $j, k, l, m \in \{1, \dots, 6\}$, it follows that

$$\|D_{\mathfrak{M}}(\Lambda_1) - D_{\mathfrak{M}}(\Lambda_2)\|_\infty \leqslant 4\,\|\Lambda_1\|_\infty \cdot \|\Lambda_1 - \Lambda_2\|_\infty + 2\,\|\Lambda_1 - \Lambda_2\|_\infty^2$$

and hence that $D_{\mathfrak{M}}$ is component-wise continuous, i.e., if $\Lambda_1, \Lambda_2, \dots$ is a component-wise to $\Lambda \in \mathcal{L}$ convergent sequence of elements of $\mathcal{L}$, then $D_{\mathfrak{M}}(\Lambda_1), D_{\mathfrak{M}}(\Lambda_2), \dots$ is a component-wise to $D_{\mathfrak{M}}(\Lambda) \in \mathcal{L}$ convergent sequence in $\mathrm{M}(6, \mathbb{R})$.

In addition, we note that since $\mathfrak{R}(S)$ is unitary if $S \in \mathrm{O}(4)$, we have that $\mathfrak{R}'(S)$ is unitary if $S \in \mathrm{O}(4)$ and hence that

if $\Lambda \in \mathcal{L} \cap \mathrm{O}(4)$, then $D_{\mathfrak{M}}(\Lambda) \in \mathrm{O}(6)$.

In particular, we have that

$$D_{\mathfrak{M}}(\Lambda_P) = D_{\mathfrak{M}}(\Lambda_T) = \begin{pmatrix} -1 & 0 & 0 & 0 & 0 & 0 \\ 0 & -1 & 0 & 0 & 0 & 0 \\ 0 & 0 & -1 & 0 & 0 & 0 \\ 0 & 0 & 0 & 1 & 0 & 0 \\ 0 & 0 & 0 & 0 & 1 & 0 \\ 0 & 0 & 0 & 0 & 0 & 1 \end{pmatrix} \in \mathrm{O}(6) ,$$

$$D_{\mathfrak{M}}(\Lambda_{PT}) = E \in \mathrm{O}(6) ,$$

where the elements Λ_T, Λ_P and Λ_{PT}, T stands for "time reversal" and P for "parity," of the Lorentz group are defined by

$$\Lambda_P := \begin{pmatrix} 1 & 0 & 0 & 0 \\ 0 & -1 & 0 & 0 \\ 0 & 0 & -1 & 0 \\ 0 & 0 & 0 & -1 \end{pmatrix} , \quad \Lambda_T := \begin{pmatrix} -1 & 0 & 0 & 0 \\ 0 & 1 & 0 & 0 \\ 0 & 0 & 1 & 0 \\ 0 & 0 & 0 & 1 \end{pmatrix} ,$$

$$\Lambda_{PT} := \Lambda_P \cdot \Lambda_T = \Lambda_T \cdot \Lambda_P = -E .$$

Further, if $M_j : \mathbb{R} \to \mathcal{L}_+^\uparrow \cap \mathrm{O}(4)$, $j \in \{1, 2, 3\}$, are the one-parameter groups of rotations about the coordinate axes defined by (1.17), then

$$D_{\mathfrak{M}}(M_1(s)) = \begin{pmatrix} 1 & 0 & 0 & 0 & 0 & 0 \\ 0 & \cos(s) & \sin(s) & 0 & 0 & 0 \\ 0 & -\sin(s) & \cos(s) & 0 & 0 & 0 \\ 0 & 0 & 0 & 1 & 0 & 0 \\ 0 & 0 & 0 & 0 & \cos(s) & \sin(s) \\ 0 & 0 & 0 & 0 & -\sin(s) & \cos(s) \end{pmatrix} \in \mathrm{O}(6) ,$$

$$D_{\mathfrak{M}}(M_2(s)) = \begin{pmatrix} \cos(s) & 0 & -\sin(s) & 0 & 0 & 0 \\ 0 & 1 & 0 & 0 & 0 & 0 \\ \sin(s) & 0 & \cos(s) & 0 & 0 & 0 \\ 0 & 0 & 0 & \cos(s) & 0 & -\sin(s) \\ 0 & 0 & 0 & 0 & 1 & 0 \\ 0 & 0 & 0 & \sin(s) & 0 & \cos(s) \end{pmatrix} \in \mathrm{O}(4) ,$$

$$D_{\mathfrak{M}}(M_3(s)) = \begin{pmatrix} \cos(s) & \sin(s) & 0 & 0 & 0 & 0 \\ -\sin(s) & \cos(s) & 0 & 0 & 0 & 0 \\ 0 & 0 & 1 & 0 & 0 & 0 \\ 0 & 0 & 0 & \cos(s) & \sin(s) & 0 \\ 0 & 0 & 0 & -\sin(s) & \cos(s) & 0 \\ 0 & 0 & 0 & 0 & 0 & 1 \end{pmatrix} \in \mathrm{O}(6) ,$$

for every $s \in \mathbb{R}$ and if $M_{01}, M_{02}, M_{03} : \mathbb{R} \to \mathcal{L}_+^\uparrow$ are the one-parameter groups of Lorentz boosts defined by (1.20), we have that

$$D_{\mathfrak{M}}(M_{01}(s)) = \begin{pmatrix} 1 & 0 & 0 & 0 & 0 & 0 \\ 0 & \cosh(s) & 0 & 0 & 0 & \sinh(s) \\ 0 & 0 & \cosh(s) & 0 & -\sinh(s) & 0 \\ 0 & 0 & 0 & 1 & 0 & 0 \\ 0 & 0 & -\sinh(s) & 0 & \cosh(s) & 0 \\ 0 & \sinh(s) & 0 & 0 & 0 & \cosh(s) \end{pmatrix} ,$$

$$D_{\mathfrak{M}}(M_{02}(s)) = \begin{pmatrix} \cosh(s) & 0 & 0 & 0 & 0 & -\sinh(s) \\ 0 & 1 & 0 & 0 & 0 & 0 \\ 0 & 0 & \cosh(s) & \sinh(s) & 0 & 0 \\ 0 & 0 & \sinh(s) & \cosh(s) & 0 & 0 \\ 0 & 0 & 0 & 0 & 1 & 0 \\ -\sinh(s) & 0 & 0 & 0 & 0 & \cosh(s) \end{pmatrix},$$

$$D_{\mathfrak{M}}(M_{03}(s)) = \begin{pmatrix} \cosh(s) & 0 & 0 & 0 & \sinh(s) & 0 \\ 0 & \cosh(s) & 0 & -\sinh(s) & 0 & 0 \\ 0 & 0 & 1 & 0 & 0 & 0 \\ 0 & -\sinh(s) & 0 & \cosh(s) & 0 & 0 \\ \sinh(s) & 0 & 0 & 0 & \cosh(s) & 0 \\ 0 & 0 & 0 & 0 & 0 & 1 \end{pmatrix},$$

for every $s \in \mathbb{R}$.

Exercise 2.5 For $j \in \{1, 2, 3\}$, $s \in \mathbb{R}$, show that

$$\|D_{\mathfrak{M}}(M_j(s)) - 1\|_\infty \leqslant |s| ,$$

for $s \in [0, 1]$.

Exercise 2.6 For $j \in \{1, 2, 3\}$, $s \in [-1, 1]$, show that

$$\|\frac{1}{s}\left[D_{\mathfrak{M}}(M_j(s)) - 1\right] - i\hat{\sigma}_j\|_\infty \leqslant \frac{|s|}{2} ,$$

where

$$\hat{\sigma}_1 := \begin{pmatrix} 0 & 0 & 0 & 0 & 0 & 0 \\ 0 & 0 & -i & 0 & 0 & 0 \\ 0 & i & 0 & 0 & 0 & 0 \\ 0 & 0 & 0 & 0 & 0 & 0 \\ 0 & 0 & 0 & 0 & 0 & -i \\ 0 & 0 & 0 & 0 & i & 0 \end{pmatrix}, \quad \hat{\sigma}_2 := \begin{pmatrix} 0 & 0 & i & 0 & 0 & 0 \\ 0 & 0 & 0 & 0 & 0 & 0 \\ -i & 0 & 0 & 0 & 0 & 0 \\ 0 & 0 & 0 & 0 & 0 & i \\ 0 & 0 & 0 & 0 & 0 & 0 \\ 0 & 0 & 0 & -i & 0 & 0 \end{pmatrix},$$

$$\hat{\sigma}_3 := \begin{pmatrix} 0 & -i & 0 & 0 & 0 & 0 \\ i & 0 & 0 & 0 & 0 & 0 \\ 0 & 0 & 0 & 0 & 0 & 0 \\ 0 & 0 & 0 & 0 & -i & 0 \\ 0 & 0 & 0 & i & 0 & 0 \\ 0 & 0 & 0 & 0 & 0 & 0 \end{pmatrix}.$$

Exercise 2.7 For $j \in \{1, 2, 3\}$, $s \in [-1, 1]$, show that

$$\|D_{\mathfrak{M}}(M_{0j}(s)) - 1\|_\infty \leqslant \sinh(|s|) \ .$$

Exercise 2.8 For $j \in \{1, 2, 3\}$, $s \in [-1, 1]$, show that

$$\|\frac{1}{is}\left[D_{\mathfrak{M}}(M_{0j}(s)) - 1\right] - \hat{\hat{\sigma}}_j\|_\infty \leqslant \sinh(|s|) \ ,$$

where

$$\hat{\hat{\sigma}}_1 := \begin{pmatrix} 0 & 0 & 0 & 0 & 0 & 0 \\ 0 & 0 & 0 & 0 & 0 & -i \\ 0 & 0 & 0 & 0 & i & 0 \\ 0 & 0 & 0 & 0 & 0 & 0 \\ 0 & 0 & i & 0 & 0 & 0 \\ 0 & -i & 0 & 0 & 0 & 0 \end{pmatrix} , \ \hat{\hat{\sigma}}_2 := \begin{pmatrix} 0 & 0 & 0 & 0 & 0 & i \\ 0 & 0 & 0 & 0 & 0 & 0 \\ 0 & 0 & 0 & -i & 0 & 0 \\ 0 & 0 & -i & 0 & 0 & 0 \\ 0 & 0 & 0 & 0 & 0 & 0 \\ i & 0 & 0 & 0 & 0 & 0 \end{pmatrix} ,$$

$$\hat{\hat{\sigma}}_3 := \begin{pmatrix} 0 & 0 & 0 & 0 & -i & 0 \\ 0 & 0 & 0 & i & 0 & 0 \\ 0 & 0 & 0 & 0 & 0 & 0 \\ 0 & i & 0 & 0 & 0 & 0 \\ -i & 0 & 0 & 0 & 0 & 0 \\ 0 & 0 & 0 & 0 & 0 & 0 \end{pmatrix} . \tag{2.2}$$

2.3 A Representation of the Poincaré Group

In the following, let $a \geqslant 0$. For $(a, \Lambda) \in \mathcal{P}$, we define $\mathcal{T}_a(a, \Lambda)$ by

$$\mathcal{T}_a(a, \Lambda) f := D_{\mathfrak{M}}(\Lambda) \cdot \begin{pmatrix} \hat{\hat{U}}_a(a, \Lambda) f_1 \\ \vdots \\ \hat{\hat{U}}_a(a, \Lambda) f_6 \end{pmatrix}$$

$$= \begin{pmatrix} \sum_{j=1}^{6} (D_{\mathfrak{M}}(\Lambda))_{1j} \hat{\hat{U}}_a(a, \Lambda) f_j \\ \vdots \\ \sum_{j=1}^{6} (D_{\mathfrak{M}}(\Lambda))_{6j} \hat{\hat{U}}_a(a, \Lambda) f_j \end{pmatrix} , \tag{2.3}$$

for every $f \in (L^2_{\mathbb{C}}(\mathbb{R}^3, \varphi_a))^6$, where $\hat{\hat{U}}_a$ is the unitary/anti-unitary representation of $\mathcal{P}$ from Part I, see Sect. 5.5 in Part I. In particular, the additive, monotone and regular interval function φ_a is defined by

$$\varphi_a(I) := \int_{\mathbb{R}^3} \chi_I \cdot (v_a^0)^{-1} \, dv^3 \ ,$$

for every bounded interval I of $\mathbb{R}^3$, where $v_a^0 : \mathbb{R}^3 \to \mathbb{R}$ is defined by

$$v_a^0(v) := (|v|^2 + a^2)^{1/2} \ ,$$

for every $v \in \mathbb{R}^3$, and v^3 denotes the Lebesgue measure on $\mathbb{R}^3$, such that

- a subset $N \subset \mathbb{R}^3$ is a φ_a-zero set if and only if it is a v^3-zero set
- and that g is φ_a-integrable if and only if $(v_a^0)^{-1} g$ is v^3-integrable and that in this case we have that

$$\int_{\mathbb{R}^3} g \, d\varphi_a = \int_{\mathbb{R}^3} g \cdot (v_a^0)^{-1} \, dv^3 \, .$$

If $(a, \Lambda) \in \mathscr{P}_+^\uparrow \cup \mathscr{P}_-^\uparrow$, the map $\mathscr{T}_a(a, \Lambda)$ is obviously linear and, if $(a, \Lambda) \in \mathscr{P}_+^\downarrow \cup \mathscr{P}_-^\downarrow$, anti-linear. Further,

$$\begin{aligned}
&\langle \mathscr{T}_a(a, \Lambda) f | \mathscr{T}_a(a, \Lambda) g \rangle \\
&= \sum_{k=1}^{6} \langle \sum_{l=1}^{6} (D_{\mathfrak{M}}(\Lambda))_{kl} \hat{\hat{U}}_a(a, \Lambda) f_l \mid \sum_{m=1}^{6} (D_{\mathfrak{M}}(\Lambda))_{km} \hat{\hat{U}}_a(a, \Lambda) g_m \rangle_2 \\
&= \sum_{k,l,m=1}^{6} [(D_{\mathfrak{M}}(\Lambda))_{kl}]^* (D_{\mathfrak{M}}(\Lambda))_{km} \langle \hat{\hat{U}}_a(a, \Lambda) f_l \, | U_a(a, \Lambda) g_m \rangle_2 \\
&= \sum_{k,l,m=1}^{6} [(D_{\mathfrak{M}}(\Lambda))_{kl}]^* (D_{\mathfrak{M}}(\Lambda))_{km} \langle f_l \, | g_m \rangle_2^{(*)} \\
&= \sum_{l,m=1}^{6} \left[\sum_{k=1}^{6} [(D_{\mathfrak{M}}(\Lambda))_{kl}]^* (D_{\mathfrak{M}}(\Lambda))_{km} \right] \langle f_l \, | g_m \rangle_2^{(*)} \\
&= \sum_{l,m=1}^{6} \left[\sum_{k=1}^{6} [(D_{\mathfrak{M}}(\Lambda))^*]_{lk} (D_{\mathfrak{M}}(\Lambda))_{km} \right] \langle f_l \, | g_m \rangle_2^{(*)} \\
&= \sum_{l,m=1}^{6} ((D_{\mathfrak{M}}(\Lambda))^* \cdot D_{\mathfrak{M}}(\Lambda))_{lm} \langle f_l \, | g_m \rangle_2^{(*)} \, ,
\end{aligned}$$

for all $f, g \in (L^2_{\mathbb{C}}(\mathbb{R}^3, \varphi_a))^6$, where here and in the following $^{(*)}$ applies to the case that $(a, \Lambda) \in \mathscr{P}_+^\downarrow \cup \mathscr{P}_-^\downarrow$, $\langle \, | \, \rangle_2$ denotes the scalar product for $L^2_{\mathbb{C}}(\mathbb{R}^3, \varphi_a)$, with induced norm $\| \; \|_2$ and the scalar product $\langle \, | \, \rangle$ on $(L^2_{\mathbb{C}}(\mathbb{R}^3, \varphi_a))^6$, with induced norm $\| \; \|$, is defined by

$$\langle f | g \rangle := \sum_{k=1}^{6} \langle f_k | g_k \rangle_2 \, ,$$

for all $f, g \in (L^2_{\mathbb{C}}(\mathbb{R}^3, \varphi_a))^2$. Hence if $\Lambda \in \mathscr{L} \cap \mathrm{O}(4)$, we have that $D_{\mathfrak{M}}(\Lambda) \in \mathrm{U}(4)$ and therefore that

$$\langle \mathscr{T}_a(a, \Lambda) f | \mathscr{T}_a(a, \Lambda) g \rangle = \sum_{l=1}^{6} \langle f_l | g_l \rangle_2^{(*)} = \langle f | g \rangle^{(*)} \, ,$$

for all $f, g \in (L^2_{\mathbb{C}}(\mathbb{R}^3, \varphi_a))^6$, i.e., if $(a, \Lambda) \in \mathscr{P}_+^\uparrow \cup \mathscr{P}_-^\uparrow$, the map $\mathscr{T}_a(a, \Lambda)$ preserves the scalar product, whereas if $(a, \Lambda) \in \mathscr{P}_+^\downarrow \cup \mathscr{P}_-^\downarrow$ it "anti-" preserves the scalar product. In the general case, we have that

$$
\begin{aligned}
\| \mathscr{T}_a(a, \Lambda) f \|^2 &= \langle \mathscr{T}_a(a, \Lambda) f | \mathscr{T}_a(a, \Lambda) f \rangle \\
&= \sum_{k,l,m=1}^{6} [(D_{\mathfrak{M}}(\Lambda))_{kl}]^* (D_{\mathfrak{M}}(\Lambda))_{km} \langle f_l \, | f_m \rangle_2^{(*)} \\
&\leqslant \| D_{\mathfrak{M}}(\Lambda) \|_\infty^2 \sum_{k,l,m=1}^{6} \| f_l \|_2 \, \| f_m \|_2 = 3 \, \| D_{\mathfrak{M}}(\Lambda) \|_\infty^2 \sum_{l,m=1}^{6} 2 \, \| f_l \|_2 \, \| f_m \|_2 \\
&\leqslant 3 \, \| D_{\mathfrak{M}}(\Lambda) \|_\infty^2 \sum_{l,m=1}^{6} (\| f_l \|_2^2 + \| f_m \|_2^2) \\
&= 3 \, \| D_{\mathfrak{M}}(\Lambda) \|_\infty^2 \sum_{l,m=1}^{6} \| f_l \|_2^2 + 3 \, \| D_{\mathfrak{M}}(\Lambda) \|_\infty^2 \sum_{l,m=1}^{6} \| f_m \|_2^2 \\
&= 36 \, \| D_{\mathfrak{M}}(\Lambda) \|_\infty^2 \, \| f \|^2 \ ,
\end{aligned}
$$

for all $f \in (L^2_{\mathbb{C}}(\mathbb{R}^3, \varphi_a))^6$, and hence that by (2.3), there is defined a map $\mathscr{T}_a$ into the union of bounded linear and bounded anti-linear maps from $(L^2_{\mathbb{C}}(\mathbb{R}^3, \varphi_a))^6$ to $(L^2_{\mathbb{C}}(\mathbb{R}^3, \varphi_a))^6$. Further, for $(a_1, \Lambda_1), (a_2, \Lambda_2) \in \mathscr{P}$ and $f \in (L^2_{\mathbb{C}}(\mathbb{R}^3, \varphi_a))^6$, we have that

$$
\begin{aligned}
\mathscr{T}_a(a_1, \Lambda_1) \mathscr{T}_a(a_2, \Lambda_2) f &= \begin{pmatrix} \sum_{j=1}^6 (D_{\mathfrak{M}}(\Lambda_1))_{1j} \hat{\hat{U}}_a(a_1, \Lambda_1) [\mathscr{T}_a(a_2, \Lambda_2) f]_j \\ \vdots \\ \sum_{j=1}^6 (D_{\mathfrak{M}}(\Lambda_1))_{6j} \hat{\hat{U}}_a(a_1, \Lambda_1) [\mathscr{T}_a(a_2, \Lambda_2) f]_j \end{pmatrix} \\
&= \begin{pmatrix} \sum_{j=1}^6 (D_{\mathfrak{M}}(\Lambda_1))_{1j} \hat{\hat{U}}_a(a_1, \Lambda_1) \sum_{k=1}^6 (D_{\mathfrak{M}}(\Lambda_2))_{jk} \hat{\hat{U}}_a(a_2, \Lambda_2) f_k \\ \vdots \\ \sum_{j=1}^6 (D_{\mathfrak{M}}(\Lambda_1))_{6j} \hat{\hat{U}}_a(a_1, \Lambda_1) \sum_{k=1}^6 (D_{\mathfrak{M}}(\Lambda_2))_{jk} \hat{\hat{U}}_a(a_2, \Lambda_2) f_k \end{pmatrix} \\
&= \begin{pmatrix} \sum_{j,k=1}^6 (D_{\mathfrak{M}}(\Lambda_1))_{1j} (D_{\mathfrak{M}}(\Lambda_2))_{jk}^{(*)} \hat{\hat{U}}_a(a_1, \Lambda_1) \hat{\hat{U}}_a(a_2, \Lambda_2) f_k \\ \vdots \\ \sum_{j,k=1}^6 (D_{\mathfrak{M}}(\Lambda_1))_{6j} (D_{\mathfrak{M}}(\Lambda_2))_{jk}^{(*)} \hat{\hat{U}}_a(a_1, \Lambda_1) \hat{\hat{U}}_a(a_2, \Lambda_2) f_k \end{pmatrix} \\
&= \begin{pmatrix} \sum_{j,k=1}^6 (D_{\mathfrak{M}}(\Lambda_1))_{1j} (D_{\mathfrak{M}}(\Lambda_2))_{jk} \hat{\hat{U}}_a(a_1, \Lambda_1) \hat{\hat{U}}_a(a_2, \Lambda_2) f_k \\ \vdots \\ \sum_{j,k=1}^6 (D_{\mathfrak{M}}(\Lambda_1))_{6j} (D_{\mathfrak{M}}(\Lambda_2))_{jk} \hat{\hat{U}}_a(a_1, \Lambda_1) \hat{\hat{U}}_a(a_2, \Lambda_2) f_k \end{pmatrix}
\end{aligned}
$$

$$= \begin{pmatrix} \sum_{k=1}^{6} (D_{\mathfrak{M}}(\Lambda_1) \cdot D_{\mathfrak{M}}(\Lambda_2))_{1k} \hat{\hat{U}}_a(a_1, \Lambda_1) \hat{\hat{U}}_a(a_2, \Lambda_2) f_k \\ \vdots \\ \sum_{k=1}^{6} (D_{\mathfrak{M}}(\Lambda_1) \cdot D_{\mathfrak{M}}(\Lambda_2))_{6k} \hat{\hat{U}}_a(a_1, \Lambda_1) \hat{\hat{U}}_a(a_2, \Lambda_2) f_k \end{pmatrix}$$

$$= \begin{pmatrix} \sum_{k=1}^{6} (D_{\mathfrak{M}}(\Lambda_1 \cdot \Lambda_2))_{1k} \hat{\hat{U}}_a((a_1, \Lambda_1) \cdot (a_2, \Lambda_2)) f_k \\ \vdots \\ \sum_{k=1}^{6} (D_{\mathfrak{M}}(\Lambda_1 \cdot \Lambda_2))_{6k} \hat{\hat{U}}_a((a_1, \Lambda_1) \cdot (a_2, \Lambda_2)) f_k \end{pmatrix}$$

$$= \mathscr{T}_a((a_1, \Lambda_1) \cdot (a_2, \Lambda_2)) f \ ,$$

where we have used that

$$(a_1, \Lambda_1) \cdot (a_2, \Lambda_2) = (a_1 + \Lambda_1 a_2, \Lambda_1 \Lambda_2) \ .$$

Since $\mathscr{T}_a(0, E)$ is given by the identity map on $(L^2_{\mathbb{C}}(\mathbb{R}^3, \varphi_a))^6$, it follows that $\mathscr{T}_a$ is a representation of $\mathscr{P}$. We note that this implies also that $\mathscr{T}_a(a, \Lambda)$ is unitary linear if $\Lambda \in (\mathscr{L}_+^\uparrow \cup \mathscr{L}_-^\uparrow) \cap \mathrm{O}(4)$ and anti-unitary anti-linear if $\Lambda \in (\mathscr{L}_+^\downarrow \cup \mathscr{L}_-^\downarrow) \cap \mathrm{O}(4)$.

Further, if $(a_1, \Lambda_1), (a_2, \Lambda_2) \in \mathscr{P}_+^\uparrow$ and $f \in (L^2_{\mathbb{C}}(\mathbb{R}^3, \varphi_a))^6$, we have that

$$\begin{aligned}
&\mathscr{T}_a(a_1, \Lambda_1) f - \mathscr{T}_a(a_2, \Lambda_2) f \\
&= D_{\mathfrak{M}}(\Lambda_1) \cdot \begin{pmatrix} \hat{\hat{U}}_a(a_1, \Lambda_1) f_1 \\ \vdots \\ \hat{\hat{U}}_a(a_1, \Lambda_1) f_6 \end{pmatrix} - D_{\mathfrak{M}}(\Lambda_2) \cdot \begin{pmatrix} \hat{\hat{U}}_a(a_2, \Lambda_2) f_1 \\ \vdots \\ \hat{\hat{U}}_a(a_2, \Lambda_2) f_6 \end{pmatrix} \\
&= D_{\mathfrak{M}}(\Lambda_1) \cdot \left[\begin{pmatrix} \hat{\hat{U}}_a(a_1, \Lambda_1) f_1 \\ \vdots \\ \hat{\hat{U}}_a(a_1, \Lambda_1) f_6 \end{pmatrix} - \begin{pmatrix} \hat{\hat{U}}_a(a_2, \Lambda_2) f_1 \\ \vdots \\ \hat{\hat{U}}_a(a_2, \Lambda_2) f_6 \end{pmatrix} \right] \\
&\quad + (D_{\mathfrak{M}}(\Lambda_1) - D_{\mathfrak{M}}(\Lambda_2)) \cdot \begin{pmatrix} \hat{\hat{U}}_a(a_2, \Lambda_2) f_1 \\ \vdots \\ \hat{\hat{U}}_a(a_2, \Lambda_2) f_6 \end{pmatrix} \\
&= D_{\mathfrak{M}}(\Lambda_1) \cdot \begin{pmatrix} [\hat{\hat{U}}_a(a_1, \Lambda_1) - \hat{\hat{U}}_a(a_2, \Lambda_2)] f_1 \\ \vdots \\ [\hat{\hat{U}}_a(a_1, \Lambda_1) - \hat{\hat{U}}_a(a_2, \Lambda_2)] f_6 \end{pmatrix} \\
&\quad + (D_{\mathfrak{M}}(\Lambda_1) - D_{\mathfrak{M}}(\Lambda_2)) \cdot \begin{pmatrix} \hat{\hat{U}}_a(a_2, \Lambda_2) f_1 \\ \vdots \\ \hat{\hat{U}}_a(a_2, \Lambda_2) f_6 \end{pmatrix}
\end{aligned}$$

$$= \begin{pmatrix} \sum_{j=1}^{6} (D_{\mathfrak{M}}(\Lambda_1))_{1j} [\hat{\hat{U}}_a(a_1, \Lambda_1) - \hat{\hat{U}}_a(a_2, \Lambda_2)] f_j \\ \vdots \\ \sum_{j=1}^{6} (D_{\mathfrak{M}}(\Lambda_1))_{6j} [\hat{\hat{U}}_a(a_1, \Lambda_1) - \hat{\hat{U}}_a(a_2, \Lambda_2)] f_j \end{pmatrix} + \begin{pmatrix} \sum_{j=1}^{6} (D_{\mathfrak{M}}(\Lambda_1) - D_{\mathfrak{M}}(\Lambda_2))_{1j} \hat{\hat{U}}_a(a_2, \Lambda_2) f_j \\ \vdots \\ \sum_{j=1}^{6} (D_{\mathfrak{M}}(\Lambda_1) - D_{\mathfrak{M}}(\Lambda_2))_{6j} \hat{\hat{U}}_a(a_2, \Lambda_2) f_j \end{pmatrix} .$$

and hence that

$$\begin{aligned} &\| \mathscr{T}_a(a_1, \Lambda_1) f - \mathscr{T}_a(a_2, \Lambda_2) f \|^2 \\ &\leqslant 2 \sum_{k,j=1}^{6} \| (D_{\mathfrak{M}}(\Lambda_1))_{kj} [\hat{\hat{U}}_a(a_1, \Lambda_1) - \hat{\hat{U}}_a(a_2, \Lambda_2)] f_j \|_2^2 \\ &\quad + 2 \sum_{k,j=1}^{6} \| (D_{\mathfrak{M}}(\Lambda_1) - D_{\mathfrak{M}}(\Lambda_2))_{kj} \hat{\hat{U}}_a(a_2, \Lambda_2) f_j \|_2^2 \\ &\leqslant 2 \sum_{k,j=1}^{6} |(D_{\mathfrak{M}}(\Lambda_1))_{kj}|^2 \cdot \| [\hat{\hat{U}}_a(a_1, \Lambda_1) - \hat{\hat{U}}_a(a_2, \Lambda_2)] f_j \|_2^2 \\ &\quad + 2 \sum_{k,j=1}^{6} |(D_{\mathfrak{M}}(\Lambda_1) - D_{\mathfrak{M}}(\Lambda_2))_{kj}| \cdot \| f_j \|_2^2 \\ &\leqslant 12 \sum_{j=1}^{6} \| D_{\mathfrak{M}}(\Lambda_1) \|_\infty^2 \cdot \| [\hat{\hat{U}}_a(a_1, \Lambda_1) - \hat{\hat{U}}_a(a_2, \Lambda_2)] f_j \|_2^2 \\ &\quad + 72 \, \| D_{\mathfrak{M}}(\Lambda_1) - D_{\mathfrak{M}}(\Lambda_2) \|_\infty \cdot \| f \|^2 . \end{aligned}$$

Since D_M is component-wise continuous and $\hat{\hat{U}}_a|_{\mathscr{P}_+^\uparrow}$ is strongly continuous, it follows that $\mathscr{T}_a$ is strongly continuous, i.e., if $(a_1, \Lambda_1), (a_2, \Lambda_2), \ldots$ is a sequence in $\mathscr{P}_+^\uparrow$ that converges component-wise to $(a, \Lambda) \in \mathscr{P}_+^\uparrow$, then

$$\lim_{\nu \to \infty} \| [\mathscr{T}_a(a_\nu, \Lambda_\nu) - \mathscr{T}_a(a, \Lambda)] f \| = 0 ,$$

for every $f \in (L^2_{\mathbb{C}}(\mathbb{R}^3, \varphi_a))^6$.

Hence, we arrive at the following result. If $a \geqslant 0$, then $\mathscr{T}_a$ defined by

$$\mathscr{T}_a(a, \Lambda)f = \begin{pmatrix} \sum_{j=1}^{6}(D_{\mathfrak{M}}(\Lambda))_{1j}\hat{\hat{U}}_a(a,\Lambda)f_j \\ \vdots \\ \sum_{j=1}^{6}(D_{\mathfrak{M}}(\Lambda))_{6j}\hat{\hat{U}}_a(a,\Lambda)f_j \end{pmatrix},$$

for every $f \in (L^2_{\mathbb{C}}(\mathbb{R}^3, \varphi_a))^6$ and every $(a, \Lambda) \in \mathscr{P}$, is a representation of $\mathscr{P}$. In addition, $\mathscr{T}_a|_{\mathscr{P}_+^\uparrow}$ is strongly continuous, i.e., if $(a_1, \Lambda_1), (a_2, \Lambda_2), \dots$ is a sequence in $\mathscr{P}_+^\uparrow$ that converges component-wise to $(a, \Lambda) \in \mathscr{P}_+^\uparrow$, then

$$\lim_{\nu \to \infty} \|[\mathscr{T}_a(a_\nu, \Lambda_\nu) - \mathscr{T}_a(a, \Lambda)]f\| = 0 ,$$

for every $f \in (L^2_{\mathbb{C}}(\mathbb{R}^3, \varphi_a))^6$. Further, if $(a, \Lambda) \in \mathscr{P}_+^\uparrow \cup \mathscr{P}_-^\uparrow$, then the map $\mathscr{T}_a(a, \Lambda)$ is linear and, if $(a, \Lambda) \in \mathscr{P}_+^\downarrow \cup \mathscr{P}_-^\downarrow$, then the map $\mathscr{T}_a(a, \Lambda)$ is anti-linear. In particular, $\mathscr{T}_a(a, \Lambda)$ is unitary linear if $\Lambda \in (\mathscr{L}_+^\uparrow \cup \mathscr{L}_-^\uparrow) \cap \mathrm{O}(4)$ and anti-unitary anti-linear if $\Lambda \in (\mathscr{L}_+^\downarrow \cup \mathscr{L}_-^\downarrow) \cap \mathrm{O}(4)$.

2.3.1 Generators Associated with One-Parameter Subgroups of $\mathscr{P}_+^\uparrow$, Rotations, Lorentz Boosts and Translations

In the following, we analyze the generators of $\mathscr{T}_a$, for one-parameter subgroups associated with some one-parameter subgroups of the Poincaré group. For this purpose, let $a \geqslant 0$.

We note that the inclusion $\iota : \mathscr{L} \hookrightarrow \mathscr{P}$, defined by $\iota(\Lambda) := (0, \Lambda)$, for every $\Lambda \in \mathscr{L}$, is a group monomorphism, since

$$\iota(\Lambda_1 \cdot \Lambda_2) = (0, \Lambda_1 \cdot \Lambda_2) = (0, \Lambda_1) \cdot (0, \Lambda_2) = \iota(\Lambda_1) \cdot \iota(\Lambda_2) ,$$

for all $\Lambda_1, \Lambda_2 \in \mathscr{L}$, and $\iota(\Lambda_1) = \iota(\Lambda_2)$, for some $\Lambda_1, \Lambda_2 \in \mathscr{L}$, implies that $\Lambda_1 = \Lambda_2$. Hence, if $M : (\mathbb{R}, +) \to \mathscr{L}$ is a one-parameter subgroup of $\mathscr{L}$, then $\iota \circ M$ is a one-parameter subgroup of $\mathscr{P}$. Further, if M is continuous, i.e., for every sequence $s_1, s_2, \dots$ in $\mathbb{R}$ that is convergent to $s \in \mathbb{R}$, the corresponding sequence $M(s_1), M(s_2), \dots$ converges component-wise to $M(s)$, then $\iota \circ M$ is continuous, i.e., for every sequence $s_1, s_2, \dots$ in $\mathbb{R}$ that is convergent to $s \in \mathbb{R}$, the corresponding sequence

$$(\iota \circ M)(s_1) = (0, M(s_1)) , \ (\iota \circ M)(s_2) = (0, M(s_2)) , \ \dots$$

converges component-wise to $(\iota \circ M)(s) = (0, M(s))$.

Then, if

$$M : (\mathbb{R}, +) \to \mathcal{L}_+^\uparrow$$

is a continuous group homomorphism, i.e., such that

$$M(s_1 + s_2) = M(s_1) \cdot M(s_2) \ ,$$

for all $s_1, s_2 \in \mathbb{R}$ and such that, for every sequence $s_1, s_2, \dots$ in $\mathbb{R}$ that is convergent to $s \in \mathbb{R}$, the corresponding sequence $M(s_1), M(s_2), \dots$ converges component-wise to $M(s)$, then $\mathcal{T}_a \circ \iota \circ M$ is a strongly continuous one-parameter group. Hence, there is an infinitesimal generator[4] A_M in $X := (L^2_{\mathbb{C}}(\mathbb{R}^3, \varphi_a))^6$, given by[5]

$$D(A_M) = \{f \in X : \lim_{s \to 0, s \neq 0} \frac{1}{s} \left[(\mathcal{T}_a \circ \iota \circ M)(s) - \mathrm{id}_X\right] f \text{ exists}\} \tag{2.4}$$

and for every $f \in D(A_M)$

$$A_M f = \frac{1}{i} \lim_{s \to 0, s \neq 0} \frac{1}{s} \left[(\mathcal{T}_a \circ \iota \circ M)(s) - \mathrm{id}_X\right] f \ . \tag{2.5}$$

2.3.1.1 Generators Associated with Rotations

In the following, we calculate the generators of one-parameter subgroups of the Poincare group, of rotations about the coordinate axes, $M_j : \mathbb{R} \to \mathcal{L}_+^\uparrow \cap \mathrm{O}(4)$, $j \in \{1, 2, 3\}$, defined by (1.17). We note that since $\mathrm{Ran}(M_j) \subset \mathrm{O}(4)$, it follows that $(\mathcal{T}_a \circ \iota \circ M_j)$ is a one-parameter unitary group and hence, according to Stone's theorem, we have that

$$\exp(isA_{M_j}) = (\mathcal{T}_a \circ \iota \circ M_j)(s) \ ,$$

for every $s \in \mathbb{R}$, where A_{M_j} is (densely-defined, linear and) self-adjoint.

For the calculation of A_{M_j}, we note that

$$(\mathcal{T}_a \circ \iota \circ M_j)(\varphi) f = D_{\mathfrak{M}}(M_j(\varphi)) \cdot \begin{pmatrix} \hat{\hat{U}}_a(\iota \circ M_j(\varphi)) f_j \\ \vdots \\ \hat{\hat{U}}_a(\iota \circ M_j(\varphi)) f_j \end{pmatrix}$$

---

[4] E.g., see [4], Sect. 4.5.

[5] The following definition is not standard. In a standard definition, the factor $1/i$ in the following definition is replaced by -1 or 1, see [4], Sect. 4.5. Hence, in the sense of strongly semigroups of operators, the generator of $\mathcal{T}_a \circ M$ is given by $-iA_M$.

$$= D_{\mathfrak{M}}(M_j(\varphi)) \cdot \begin{pmatrix} U_a(M_j(\varphi)) f_j \\ \vdots \\ U_a(M_j(\varphi)) f_j \end{pmatrix} = D_{\mathfrak{M}}(M_j(\varphi)) \cdot \begin{pmatrix} \exp\left(\frac{i\varphi}{\hbar} \hat{L}_j\right) f_1 \\ \vdots \\ \exp\left(\frac{i\varphi}{\hbar} \hat{L}_j\right) f_6 \end{pmatrix},$$

for every $\varphi \in \mathbb{R}$ and $f \in X$, where Formula 5.19 of Part I has been used. Further, for $f \in X$ and $\varphi \in \mathbb{R}^*$, it follows that

$$\begin{aligned}
&\frac{1}{i\varphi}\left[(\mathscr{T}_a \circ \iota \circ M_j)(\varphi) - \mathrm{id}_X\right] f \\
&= \frac{1}{i\varphi}\left[D_{\mathfrak{M}}(M_j(\varphi)) \cdot \begin{pmatrix} \exp\left(\frac{i\varphi}{\hbar} \hat{L}_j\right) f_1 \\ \vdots \\ \exp\left(\frac{i\varphi}{\hbar} \hat{L}_j\right) f_6 \end{pmatrix} - \begin{pmatrix} f_1 \\ \vdots \\ f_6 \end{pmatrix} \right] \\
&= D_{\mathfrak{M}}(M_j(\varphi)) \cdot \begin{pmatrix} (i\varphi)^{-1}[\exp\left(\frac{i\varphi}{\hbar} \hat{L}_j\right) - 1] f_1 \\ \vdots \\ (i\varphi)^{-1}[\exp\left(\frac{i\varphi}{\hbar} \hat{L}_j\right) - 1] f_6 \end{pmatrix} \\
&\quad + \frac{1}{i\varphi}[D_{\mathfrak{M}}(M_j(\varphi)) - 1] \begin{pmatrix} f_1 \\ \vdots \\ f_6 \end{pmatrix} \\
&= \begin{pmatrix} (i\varphi)^{-1}[\exp\left(\frac{i\varphi}{\hbar} \hat{L}_j\right) - 1] f_1 \\ \vdots \\ (i\varphi)^{-1}[\exp\left(\frac{i\varphi}{\hbar} \hat{L}_j\right) - 1] f_6 \end{pmatrix} + \hat{\sigma}_j \begin{pmatrix} f_1 \\ \vdots \\ f_6 \end{pmatrix} \\
&\quad + [D_{\mathfrak{M}}(M_j(\varphi)) - 1] \cdot \begin{pmatrix} (i\varphi)^{-1}[\exp\left(\frac{i\varphi}{\hbar} \hat{L}_j\right) - 1] f_1 \\ \vdots \\ (i\varphi)^{-1}[\exp\left(\frac{i\varphi}{\hbar} \hat{L}_j\right) - 1] f_6 \end{pmatrix} \\
&\quad + \left\{\frac{1}{i\varphi}[D_{\mathfrak{M}}(M_j(\varphi)) - 1] - \hat{\sigma}_j\right\} \begin{pmatrix} f_1 \\ \vdots \\ f_6 \end{pmatrix},
\end{aligned}$$

where the elements $\hat{\sigma}_1, \hat{\sigma}_2$ and $\hat{\sigma}_3$ of $\mathrm{M}(6, \mathbb{C})$ are given in Exercise 2.6. Hence, it follows with the help of Exercises 2.5 and 2.6 that an element $f \in X$ is part of the domain $D(A_{M_j})$ of A_{M_j} if and only if $f \in (D(\hat{L}_j))^6$ and if $f \in D(A_{M_j})$, then

$$A_{M_j} f = \frac{1}{\hbar} \begin{pmatrix} \hat{L}_j f_1 \\ \vdots \\ \hat{L}_j f_6 \end{pmatrix} + \hat{\sigma}_j \cdot \begin{pmatrix} f_1 \\ \vdots \\ f_6 \end{pmatrix}.$$

Then, according to Stone's theorem,

$$\exp(isA_G^{\hbar_R}) = (T_a^{\hbar_R} \circ G)(s) ,$$

for every $s \in \mathbb{R}$, where $A_G^{\hbar_R}$ is (densely-defined, linear and) self-adjoint.

Hence, for $j \in \{1, 2, 3\}$, we define the, densely-defined, linear and self-adjoint, j-th component $\hat{J}_j : (D(\hat{L}_j))^6 \to X$ of total angular momentum by

$$\hat{J}_j f := \hbar A_{M_j} f = \begin{pmatrix} \hat{L}_j f_1 \\ \vdots \\ \hat{L}_j f_6 \end{pmatrix} + \hbar \hat{\sigma}_j \cdot \begin{pmatrix} f_1 \\ \vdots \\ f_6 \end{pmatrix} , \tag{2.6}$$

for every $f \in (D(\hat{L}_j))^6$ as well as the, bounded linear and self-adjoint, j-th component $\hat{S}_j : X \to X$ of intrinsic angular momentum by

$$\hat{S}_j f := \hbar \hat{\sigma}_j \cdot f ,$$

for every $f \in X$, where $X = (L^2_{\mathbb{C}}(\mathbb{R}^3, \varphi_a))^6$ and

$$\hat{\sigma}_1 := \begin{pmatrix} 0 & 0 & 0 & 0 & 0 & 0 \\ 0 & 0 & -i & 0 & 0 & 0 \\ 0 & i & 0 & 0 & 0 & 0 \\ 0 & 0 & 0 & 0 & 0 & 0 \\ 0 & 0 & 0 & 0 & 0 & -i \\ 0 & 0 & 0 & 0 & i & 0 \end{pmatrix} , \quad \hat{\sigma}_2 := \begin{pmatrix} 0 & 0 & i & 0 & 0 & 0 \\ 0 & 0 & 0 & 0 & 0 & 0 \\ -i & 0 & 0 & 0 & 0 & 0 \\ 0 & 0 & 0 & 0 & 0 & i \\ 0 & 0 & 0 & 0 & 0 & 0 \\ 0 & 0 & 0 & -i & 0 & 0 \end{pmatrix} ,$$

$$\hat{\sigma}_3 := \begin{pmatrix} 0 & -i & 0 & 0 & 0 & 0 \\ i & 0 & 0 & 0 & 0 & 0 \\ 0 & 0 & 0 & 0 & 0 & 0 \\ 0 & 0 & 0 & 0 & -i & 0 \\ 0 & 0 & 0 & i & 0 & 0 \\ 0 & 0 & 0 & 0 & 0 & 0 \end{pmatrix} .$$

Further, we note that the eigenvalues of $\hat{S}_j$ are given by $-\hbar, 0$ and $\hbar$ and that there is a Hilbert basis of corresponding eigenvectors. In particular, we have that

$$\ker(\hat{S}_1 - \hbar) = L^2_{\mathbb{C}}(\mathbb{R}^3, \varphi_a) . \begin{pmatrix} 0 \\ 0 \\ 0 \\ 0 \\ -i \\ 1 \end{pmatrix} + L^2_{\mathbb{C}}(\mathbb{R}^3, \varphi_a) . \begin{pmatrix} 0 \\ -i \\ 1 \\ 0 \\ 0 \\ 0 \end{pmatrix} ,$$

$$\ker(\hat{S}_1) = L^2_{\mathbb{C}}(\mathbb{R}^3, \varphi_a) \, . \begin{pmatrix} 0 \\ 0 \\ 0 \\ 1 \\ 0 \\ 0 \end{pmatrix} + L^2_{\mathbb{C}}(\mathbb{R}^3, \varphi_a) \, . \begin{pmatrix} 1 \\ 0 \\ 0 \\ 0 \\ 0 \\ 0 \end{pmatrix},$$

$$\ker(\hat{S}_1 + \hbar) = L^2_{\mathbb{C}}(\mathbb{R}^3, \varphi_a) \, . \begin{pmatrix} 0 \\ 0 \\ 0 \\ 0 \\ i \\ 1 \end{pmatrix} + L^2_{\mathbb{C}}(\mathbb{R}^3, \varphi_a) \, . \begin{pmatrix} 0 \\ i \\ 1 \\ 0 \\ 0 \\ 0 \end{pmatrix},$$

$$\ker(\hat{S}_2 - \hbar) = L^2_{\mathbb{C}}(\mathbb{R}^3, \varphi_a) \, . \begin{pmatrix} 0 \\ 0 \\ 0 \\ i \\ 0 \\ 1 \end{pmatrix} + L^2_{\mathbb{C}}(\mathbb{R}^3, \varphi_a) \, . \begin{pmatrix} i \\ 0 \\ 1 \\ 0 \\ 0 \\ 0 \end{pmatrix},$$

$$\ker(\hat{S}_2) = L^2_{\mathbb{C}}(\mathbb{R}^3, \varphi_a) \, . \begin{pmatrix} 0 \\ 0 \\ 0 \\ 0 \\ 1 \\ 0 \end{pmatrix} + L^2_{\mathbb{C}}(\mathbb{R}^3, \varphi_a) \, . \begin{pmatrix} 0 \\ 1 \\ 0 \\ 0 \\ 0 \\ 0 \end{pmatrix},$$

$$\ker(\hat{S}_2 + \hbar) = L^2_{\mathbb{C}}(\mathbb{R}^3, \varphi_a) \, . \begin{pmatrix} 0 \\ 0 \\ 0 \\ -i \\ 0 \\ 1 \end{pmatrix} + L^2_{\mathbb{C}}(\mathbb{R}^3, \varphi_a) \, . \begin{pmatrix} -i \\ 0 \\ 1 \\ 0 \\ 0 \\ 0 \end{pmatrix},$$

$$\ker(\hat{S}_3 - \hbar) = L^2_{\mathbb{C}}(\mathbb{R}^3, \varphi_a) \, . \begin{pmatrix} 0 \\ 0 \\ 0 \\ -i \\ 1 \\ 0 \end{pmatrix} + L^2_{\mathbb{C}}(\mathbb{R}^3, \varphi_a) \, . \begin{pmatrix} -i \\ 1 \\ 0 \\ 0 \\ 0 \\ 0 \end{pmatrix},$$

$$\ker(\hat{S}_3) = L^2_{\mathbb{C}}(\mathbb{R}^3, \varphi_a) \, . \begin{pmatrix} 0 \\ 0 \\ 0 \\ 0 \\ 0 \\ 1 \end{pmatrix} + L^2_{\mathbb{C}}(\mathbb{R}^3, \varphi_a) \, . \begin{pmatrix} 0 \\ 0 \\ 1 \\ 0 \\ 0 \\ 0 \end{pmatrix},$$

$$\ker(\hat{S}_3 + \hbar) = L^2_{\mathbb{C}}(\mathbb{R}^3, \varphi_a) \, . \begin{pmatrix} 0 \\ 0 \\ 0 \\ i \\ 1 \\ 0 \end{pmatrix} + L^2_{\mathbb{C}}(\mathbb{R}^3, \varphi_a) \, . \begin{pmatrix} i \\ 1 \\ 0 \\ 0 \\ 0 \\ 0 \end{pmatrix},$$

Hence, we obtain the following result.

For $j \in \{1, 2, 3\}$, there is a Hilbert basis of X, consisting of eigenvectors of $\hat{S}_j$, corresponding to the eigenvalues $-\hbar$, 0 and $\hbar$. Therefore, $\hat{S}_j$ has a pure point spectrum given by

$$\sigma(\hat{S}_j) = \{-\hbar, 0, \hbar\} \ .$$

2.3.1.2 Generators Associated with Lorentz Boosts

In the following, we calculate the generators of one-parameter subgroups of the Poincare group, of Lorentz boosts, $M_{0j} : \mathbb{R} \to \mathscr{L}_+^\uparrow$, $j \in \{1, 2, 3\}$, defined by (1.20),

For the calculation of $A_{M_{0j}}$, we note that

$$(\mathscr{T}_a \circ \iota \circ M_{0j})(s) f = D_{\mathfrak{M}}(M_{0j}(s)) \cdot \begin{pmatrix} \hat{\hat{U}}_a(\iota \circ M_{0j}(s)) f_j \\ \vdots \\ \hat{\hat{U}}_a(\iota \circ M_{0j}(s)) f_j \end{pmatrix}$$

$$= D_{\mathfrak{M}}(M_{0j}(s)) \cdot \begin{pmatrix} U_a(M_{0j}(s)) f_j \\ \vdots \\ U_a(M_{0j}(s)) f_j \end{pmatrix} = D_{\mathfrak{M}}(M_{0j}(s)) \cdot \begin{pmatrix} \exp\left(\frac{is}{\hbar} \hat{L}_{0j}\right) f_1 \\ \vdots \\ \exp\left(\frac{is}{\hbar} \hat{L}_{0j}\right) f_6 \end{pmatrix} ,$$

for every $s \in \mathbb{R}$ and $f \in X$, where Formula 5.21 of Part I has been used. Further, for $f \in X$ and $s \in \mathbb{R}^*$, it follows that

$$\frac{1}{is} \left[(\mathscr{T}_a \circ \iota \circ M_{0j})(s) - \mathrm{id}_X \right] f$$

$$= \frac{1}{is} \left[D_{\mathfrak{M}}(M_{0j}(s)) \cdot \begin{pmatrix} \exp\left(\frac{is}{\hbar} \hat{L}_{0j}\right) f_1 \\ \vdots \\ \exp\left(\frac{is}{\hbar} \hat{L}_{0j}\right) f_6 \end{pmatrix} - \begin{pmatrix} f_1 \\ \vdots \\ f_6 \end{pmatrix} \right]$$

$$= D_{\mathfrak{M}}(M_{0j}(s)) \cdot \begin{pmatrix} (is)^{-1} [\exp\left(\frac{is}{\hbar} \hat{L}_{0j}\right) - 1] f_1 \\ \vdots \\ (is)^{-1} [\exp\left(\frac{is}{\hbar} \hat{L}_{0j}\right) - 1] f_6 \end{pmatrix}$$

$$+ \frac{1}{is} [D_{\mathfrak{M}}(M_{0j}(s)) - 1] \begin{pmatrix} f_1 \\ \vdots \\ f_6 \end{pmatrix}$$

$$= \begin{pmatrix} (is)^{-1}[\exp\left(\frac{is}{\hbar}\,\hat{L}_{0j}\right) - 1]f_1 \\ \vdots \\ (is)^{-1}[\exp\left(\frac{is}{\hbar}\,\hat{L}_{0j}\right) - 1]f_6 \end{pmatrix} + \hat{\hat{\sigma}}_j \begin{pmatrix} f_1 \\ \vdots \\ f_6 \end{pmatrix}$$

$$+ [D_{\mathfrak{M}}(M_{0j}(s)) - 1] \cdot \begin{pmatrix} (is)^{-1}[\exp\left(\frac{is}{\hbar}\,\hat{L}_{0j}\right) - 1]f_1 \\ \vdots \\ (is)^{-1}[\exp\left(\frac{is}{\hbar}\,\hat{L}_{0j}\right) - 1]f_6 \end{pmatrix}$$

$$+ \left\{ \frac{1}{is}\,[D_{\mathfrak{M}}(M_{0j}(s)) - 1] - \hat{\hat{\sigma}}_j \right\} \begin{pmatrix} f_1 \\ \vdots \\ f_6 \end{pmatrix},$$

where the elements $\hat{\hat{\sigma}}_1, \hat{\hat{\sigma}}_2$ and $\hat{\hat{\sigma}}_3$ of $\mathrm{M}(6, \mathbb{C})$ are given in Exercise 2.8. Hence, it follows with the help of Exercises 2.7 and 2.8 that an element $f \in X$ is part of the domain $D(A_{M_{0j}})$ of $A_{M_{0j}}$ if and only if $f \in (D(\hat{L}_{0j}))^6$ and if $f \in D(A_{M_{0j}})$, then

$$A_{M_{0j}} f = \frac{1}{\hbar} \begin{pmatrix} \hat{L}_{0j} f_1 \\ \vdots \\ \hat{L}_{0j} f_6 \end{pmatrix} + \hat{\hat{\sigma}}_j \cdot \begin{pmatrix} f_1 \\ \vdots \\ f_6 \end{pmatrix}.$$

Hence, for $j \in \{1, 2, 3\}$, by

$$\hat{J}_{0j} f := \hbar A_{M_{0j}} f = \begin{pmatrix} \hat{L}_{0j} f_1 \\ \vdots \\ \hat{L}_{0j} f_6 \end{pmatrix} + \hbar \hat{\hat{\sigma}}_j \cdot \begin{pmatrix} f_1 \\ \vdots \\ f_6 \end{pmatrix}, \tag{2.7}$$

for every $f \in (D(\hat{L}_{0j}))^6$, where

$$\hat{\hat{\sigma}}_1 := \begin{pmatrix} 0 & 0 & 0 & 0 & 0 & 0 \\ 0 & 0 & 0 & 0 & 0 & -i \\ 0 & 0 & 0 & 0 & i & 0 \\ 0 & 0 & 0 & 0 & 0 & 0 \\ 0 & 0 & i & 0 & 0 & 0 \\ 0 & -i & 0 & 0 & 0 & 0 \end{pmatrix}, \quad \hat{\hat{\sigma}}_2 := \begin{pmatrix} 0 & 0 & 0 & 0 & 0 & i \\ 0 & 0 & 0 & 0 & 0 & 0 \\ 0 & 0 & 0 & -i & 0 & 0 \\ 0 & 0 & -i & 0 & 0 & 0 \\ 0 & 0 & 0 & 0 & 0 & 0 \\ i & 0 & 0 & 0 & 0 & 0 \end{pmatrix},$$

$$\hat{\hat{\sigma}}_3 := \begin{pmatrix} 0 & 0 & 0 & 0 & -i & 0 \\ 0 & 0 & 0 & i & 0 & 0 \\ 0 & 0 & 0 & 0 & 0 & 0 \\ 0 & i & 0 & 0 & 0 & 0 \\ -i & 0 & 0 & 0 & 0 & 0 \\ 0 & 0 & 0 & 0 & 0 & 0 \end{pmatrix},$$

there is given a densely-defined, linear operator $\hat{J}_{0j} : (D(\hat{L}_{0j}))^6 \to X$ in $(L^2_{\mathbb{C}}(\mathbb{R}^3, \varphi_a))^6$, such that $-(i/\hbar)\hat{J}_{0j}$ is the generator of a strongly continuous one-parameter group on $(L^2_{\mathbb{C}}(\mathbb{R}^3, \varphi_a))^6$.

2.3.1.3 Generators Associated with Translations

In the following, we calculate the generators of one-parameter subgroups of the Poincare group, of translations $M : \mathbb{R} \to \mathcal{P}$, where

$$M(s) := (sa, E) \ ,$$

for every $s \in \mathbb{R}$ and $a \in \mathbb{R}^4$. The map M is a one-parameter group of $\mathcal{P}$, since

$$\begin{aligned} M(0) &= (0, E) \ , \\ M(s_1) \cdot M(s_2) &= (s_1 a, E) \cdot (s_2 a, E) \\ &= (s_1 a + E\, s_2 a, E \cdot E) \\ &= ((s_1 + s_2)\, a, E) = M(s_1 + s_2) \ , \end{aligned}$$

for all $s_1, s_2 \in \mathbb{R}$. In addition, M is component-wise continuous. Further, for every $s \in \mathbb{R}$ and $f \in (L^2_{\mathbb{C}}(\mathbb{R}^3, \varphi_a))^6$, we have that

$$\begin{aligned} (\mathcal{T}_a \circ M)(s) f &= D_{\mathfrak{M}}(E) \cdot \begin{pmatrix} \hat{\hat{U}}_a(sa, E) f_1 \\ \vdots \\ \hat{\hat{U}}_a(sa, E) f_6 \end{pmatrix} = \begin{pmatrix} \hat{U}_a(sa, E) f_1 \\ \vdots \\ \hat{U}_a(sa, E) f_6 \end{pmatrix} \\ &= \begin{pmatrix} \exp(is\, T_{a_0 v_a^0 - \vec{a} \cdot \mathrm{id}_{\mathbb{R}^3}}) f_1 \\ \vdots \\ \exp(is\, T_{a_0 v_a^0 - \vec{a} \cdot \mathrm{id}_{\mathbb{R}^3}}) f_6 \end{pmatrix} , \end{aligned}$$

where we used the representation of $\hat{U}_a$ from Part I, see Sect. 5.4.1 of Part I, and where $T_{a_0 v_a^0 - \vec{a} \cdot \mathrm{id}_{\mathbb{R}^3}}$ denotes the maximal multiplication operator in $L^2_{\mathbb{C}}(\mathbb{R}^3, \varphi_a)$ corresponding to the function $a_0 v_a^0 - \vec{a} \cdot \mathrm{id}_{\mathbb{R}^3}$, $\vec{a} := {}^t(a_1, a_2, a_3)$ $\vec{a} := {}^t(a_1, a_2, a_3)$, $\vec{a} \cdot \mathrm{id}_{\mathbb{R}^3} : \mathbb{R}^3 \to \mathbb{R}^3$ is defined by $(\vec{a} \cdot \mathrm{id}_{\mathbb{R}^3})(v) := \vec{a} \cdot v$, for every $v \in \mathbb{R}^3$.

Hence, $\mathcal{T}_a \circ M$ is a strongly continuous one-parameter unitary group. According to Stone's theorem, there is a unique densely-defined, linear and self-adjoint operator A_M in $X := (L^2_{\mathbb{C}}(\mathbb{R}^3, \varphi_a))^6$ such that

$$\exp(isA_M) = (\mathcal{T}_a \circ M)(s) \ ,$$

for every $s \in \mathbb{R}$ and, in particular, that $A_M : D(A_M) \to X$ is given by

$$D(A_M) = \{f \in X : \lim_{s \to 0, s \neq 0} \frac{1}{s} \left[(\mathcal{T}_a \circ M)(s) - \mathrm{id}_X\right] f \text{ exists}\}$$

and for every $f \in D(A_M)$

$$A_M f = \frac{1}{i} \lim_{s \to 0, s \neq 0} \frac{1}{s} \left[(\mathcal{T}_a \circ M)(s) - \mathrm{id}_X\right] f \ .$$

Further, for $f \in X$ and $s \in \mathbb{R}^*$, it follows that

$$\frac{1}{is} \left[(\mathcal{T}_a \circ M)(s) - \mathrm{id}_X\right] f = \begin{pmatrix} \frac{1}{is} \left[\exp(is\, T_{a_0 v_a^0 - \vec{a} \cdot \mathrm{id}_{\mathbb{R}^3}}) - 1\right] f_1 \\ \vdots \\ \frac{1}{is} \left[\exp(is\, T_{a_0 v_a^0 - \vec{a} \cdot \mathrm{id}_{\mathbb{R}^3}}) - 1\right] f_6 \end{pmatrix} .$$

We note that the coordinate projections $p_1, \dots p_6 : X \to L^2_{\mathbb{C}}(\mathbb{R}^3, \varphi_a)$ as well as the inclusions $\iota_1, \dots \iota_6 : L^2_{\mathbb{C}}(\mathbb{R}^3, \varphi_a) \hookrightarrow X$, for $k \in \{1, \dots, 6\}$, defined by $p_k f := f_k$, for every $f \in X$ and, for every $f \in L^2_{\mathbb{C}}(\mathbb{R}^3, \varphi_a)$, $\iota_k f$ given by the element of X whose k-th component is given by f, whereas all other components vanish, are linear and continuous, since

$$\|p_k f\|_2 = \|f_k\|_2 \leqslant \|f\| \ , \ \|\iota_k g\| = \|g\|_2 \ ,$$

for every $f \in X$, $g \in L^2_{\mathbb{C}}(\mathbb{R}^3, \varphi_a)$ and $k \in \{1, \dots, 6\}$. Hence, $f \in D(A_M)$ if and only if f is part of the domain of

$$\bigtimes_{k=1}^{6} T_{a_0 v_a^0 - \vec{a} \cdot \mathrm{id}_{\mathbb{R}^3}}$$

and if $f \in D(A_M)$, then

$$A_M f = \left(\bigtimes_{j=1}^{6} T_{a_0 v_a^0 - \vec{a} \cdot \mathrm{id}_{\mathbb{R}^3}} \right) f \ .$$

Hence, we arrive at the following result.

Generators Associated with Translations in $(L^2_{\mathbb{C}}(\mathbb{R}^3, \varphi_a))^6$

For $a \in \mathbb{R}^4$, we have that

$$(\mathcal{T}_a \circ M)(s) = \exp\left(is \bigtimes_{k=1}^{6} T_{a_0 v_a^0 - \vec{a} \cdot \mathrm{id}_{\mathbb{R}^3}} \right) ,$$

for every $s \in \mathbb{R}$, where the one-parameter subgroup of $\mathcal{P}$, $M : \mathbb{R} \to \mathcal{P}$ is defined by $M(s) := (sa, E)$, for every $s \in \mathbb{R}$, $T_{a_0 v_a^0 - \vec{a} \cdot \mathrm{id}_{\mathbb{R}^3}}$ denotes the maximal multiplication operator in $L^2_{\mathbb{C}}(\mathbb{R}^3, \varphi_a)$, corresponding to the function $a_0 v_a^0 - \vec{a} \cdot \mathrm{id}_{\mathbb{R}^3}$, $\vec{a} := {}^t(a_1, a_2, a_3)$ and $\vec{a} \cdot \mathrm{id}_{\mathbb{R}^3} : \mathbb{R}^3 \to \mathbb{R}^3$ is defined by $(\vec{a} \cdot \mathrm{id}_{\mathbb{R}^3})(v) := \vec{a} \cdot v$, for every $v \in \mathbb{R}^3$.

Exercise 2.9 Show that the spectrum of $\bigtimes_{k=1}^{6} T_{a_0 v_a^0 - \vec{a} \cdot \mathrm{id}_{\mathbb{R}^3}}$ is given by the closure of the range of $a_0 v_a^0 - \vec{a} \cdot \mathrm{id}_{\mathbb{R}^3}$

$$\overline{\mathrm{Ran}(a_0 v_a^0 - \vec{a} \cdot \mathrm{id}_{\mathbb{R}^3})} \ .$$

Exercise 2.10 Show that, if $a \in \mathbb{R}^4$ is future-oriented, timelike, and, in particular, such that $a \cdot a = 1$, the spectrum of $\bigtimes_{k=1}^{6} T_{a_0 v_a^0 - \vec{a} \cdot \mathrm{id}_{\mathbb{R}^3}}$ is given by the closed interval $[a, \infty)$.

As a side remark, using the Hilbert space isomorphism $\mathcal{V}_a : L^2_{\mathbb{C}}(\mathbb{R}^3, \varphi_a) \to L^2_{\mathbb{C}}(\mathbb{R}^3)$ from Exercise 5.17 of Part I, defined by

$$\mathcal{V}_a f := (v_a^0)^{-1/2} f \ ,$$

for every $f \in L^2_{\mathbb{C}}(\mathbb{R}^3, \varphi_a)$, with inverse $\mathcal{V}_a^{-1} : L^2_{\mathbb{C}}(\mathbb{R}^3) \to L^2_{\mathbb{C}}(\mathbb{R}^3, \varphi_a)$, given by

$$\mathcal{V}_a^{-1} f := (v_a^0)^{1/2} f \ ,$$

for every $f \in L^2_{\mathbb{C}}(\mathbb{R}^3)$, we note that

$$\bigtimes_{k=1}^{6} \mathcal{V}_a : (L^2_{\mathbb{C}}(\mathbb{R}^3, \varphi_a))^6 \to (L^2_{\mathbb{C}}(\mathbb{R}^3))^6$$

is a Hilbert space isomorphism and that

$$\left(\bigtimes_{k=1}^{6} \mathcal{V}_a \right) \mathcal{T}_a(M(s)) \left(\bigtimes_{k=1}^{6} \mathcal{V}_a \right)^{-1} f = \exp\left(is \bigtimes_{k=1}^{6} T_{a_0 v_a^0 - \vec{a} \cdot \mathrm{id}_{\mathbb{R}^3}} \right) f \ ,$$

for every $s \in \mathbb{R}$ and $f \in (L^2_{\mathbb{C}}(\mathbb{R}^3))^6$, where $T_{a_0 v_a^0 - \vec{a} \cdot \mathrm{id}_{\mathbb{R}^3}}$ denotes the maximal multiplication operator in $L^2_{\mathbb{C}}(\mathbb{R}^3)$, corresponding to the function $a_0 v_a^0 - \vec{a} \cdot \mathrm{id}_{\mathbb{R}^3}$, $\vec{a} := {}^t(a_1, a_2, a_3)$. As a consequence, we also have the following.

Generators Associated with Translations in $(L^2_{\mathbb{C}}(\mathbb{R}^3))^6$
For $a \in \mathbb{R}^4$, we have that

$$\left(\bigtimes_{k=1}^{6} \mathcal{V}_a\right) \mathcal{T}_a(M(s)) \left(\bigtimes_{k=1}^{6} \mathcal{V}_a\right)^{-1} f = \exp\left(is \bigtimes_{k=1}^{6} T_{a_0 v_a^0 - \vec{a} \cdot \mathrm{id}_{\mathbb{R}^3}}\right) f ,$$

for every $s \in \mathbb{R}$, where the one-parameter subgroup $M : \mathbb{R} \to \mathcal{P}$ of $\mathcal{P}$, is defined by $M(s) := (sa, E)$, for every $s \in \mathbb{R}$, $T_{a_0 v_a^0 - \vec{a} \cdot \mathrm{id}_{\mathbb{R}^3}}$ denotes the maximal multiplication operator in $L^2_{\mathbb{C}}(\mathbb{R}^3)$, corresponding to the function $a_0 v_a^0 - \vec{a} \cdot \mathrm{id}_{\mathbb{R}^3}$, $\vec{a} := {}^t(a_1, a_2, a_3)$ and $\vec{a} \cdot \mathrm{id}_{\mathbb{R}^3} : \mathbb{R}^3 \to \mathbb{R}^3$ is defined by $(\vec{a} \cdot \mathrm{id}_{\mathbb{R}^3})(v) := \vec{a} \cdot v$, for every $v \in \mathbb{R}^3$.

2.4 Maxwell's Equations

Maxwell's equations for the electric field E and the magnetic field B in Minkowski space and inertial coordinates t, x, y, z are given by

$$\frac{\partial E}{\partial t} = c \, \nabla \times B \; , \; \frac{\partial B}{\partial t} = -c \, \nabla \times E \; ,$$
$$\nabla \cdot E = 0 \; , \; \nabla \cdot B = 0 \; , \tag{2.8}$$

where c denotes the speed of light. For simplicity, in the following, we allow E and B to assume complex values and assume E and B as maps on $\mathbb{R}^4$ that are everywhere partially differentiable. Through the definition

$$u_0 := \kappa c t \; , \; u_1 := \kappa x \; , \; u_2 := \kappa y \; , \; u_3 := \kappa z \; ,$$

we arrive at dimensionless coordinate projections $u_k : \mathbb{R}^4 \to \mathbb{R}$, $k \in \{0, 1, 2, 3\}$, where $\kappa > 0$ is a scale factor with dimension $1/\text{length}$. In the next step, we define the dimensionless as well as everywhere partially differentiable map $\psi : \mathbb{R}^4 \to \mathbb{C}^6$ by

$$\psi(u) := \frac{1}{\mathcal{E}_0} \begin{pmatrix} E_x((\kappa c)^{-1} u_0, \kappa^{-1} u_1, \kappa^{-1} u_2, \kappa^{-1} u_3) \\ E_y((\kappa c)^{-1} u_0, \kappa^{-1} u_1, \kappa^{-1} u_2, \kappa^{-1} u_3) \\ E_z((\kappa c)^{-1} u_0, \kappa^{-1} u_1, \kappa^{-1} u_2, \kappa^{-1} u_3) \\ B_x((\kappa c)^{-1} u_0, \kappa^{-1} u_1, \kappa^{-1} u_2, \kappa^{-1} u_3) \\ B_y((\kappa c)^{-1} u_0, \kappa^{-1} u_1, \kappa^{-1} u_2, \kappa^{-1} u_3) \\ B_z((\kappa c)^{-1} u_0, \kappa^{-1} u_1, \kappa^{-1} u_2, \kappa^{-1} u_3) \end{pmatrix} ,$$

for every $u = (u_0, u_1, u_2, u_3) \in \mathbb{R}^4$, where $\mathcal{E}_0 > 0$ has the dimension of an electric and hence also magnetic field strength $\mathrm{m}^{1/2}\,\mathrm{l}^{-1/2}\,\mathrm{t}^{-1}$. Therefore, for every $(t, x, y, z) \in \mathbb{R}^4$, we have that

$$\mathcal{E}_0 \psi(\kappa ct, \kappa x, \kappa y, \kappa z) = \begin{pmatrix} E_x(t,x,y,z) \\ E_y(t,x,y,z) \\ E_z(t,x,y,z) \\ B_x(t,x,y,z) \\ B_y(t,x,y,z) \\ B_z(t,x,y,z) \end{pmatrix}$$

and that

$$\frac{\partial E_x}{\partial t}(t,x,y,z) = \mathcal{E}_0\,\kappa c\,\frac{\partial \psi_1}{\partial u_0}(\kappa ct, \kappa x, \kappa y, \kappa z)\ ,$$
$$\frac{\partial E_x}{\partial x}(t,x,y,z) = \mathcal{E}_0\,\kappa\,\frac{\partial \psi_1}{\partial u_1}(\kappa ct, \kappa x, \kappa y, \kappa z)\ ,$$

as well for $u \in \mathbb{R}^4$ that

$$\frac{\partial E_x}{\partial t}((\kappa c)^{-1}u_0, \kappa^{-1}u_1, \kappa^{-1}u_2, \kappa^{-1}u_3) = \mathcal{E}_0\,\kappa c\,\frac{\partial \psi_1}{\partial u_0}(u)\ ,$$
$$\frac{\partial E_x}{\partial x}((\kappa c)^{-1}u_0, \kappa^{-1}u_1, \kappa^{-1}u_2, \kappa^{-1}u_3) = \mathcal{E}_0\,\kappa\,\frac{\partial \psi_1}{\partial u_1}(u)\ .$$

Therefore it follows for $u \in \mathbb{R}^4$ that

$$\mathcal{E}_0\,\kappa c\,\frac{\partial \psi_{123}}{\partial u_0}(u) = \frac{\partial E}{\partial t}((\kappa c)^{-1}u_0, \kappa^{-1}u_1, \kappa^{-1}u_2, \kappa^{-1}u_3)$$
$$= c\ (\nabla \times B)((\kappa c)^{-1}u_0, \kappa^{-1}u_1, \kappa^{-1}u_2, \kappa^{-1}u_3) = \mathcal{E}_0\,\kappa c\,(\nabla \times \psi_{456})(u)\ ,$$
$$\mathcal{E}_0\,\kappa c\,\frac{\partial \psi_{456}}{\partial u_0}(u) = \frac{\partial B}{\partial t}((\kappa c)^{-1}u_0, \kappa^{-1}u_1, \kappa^{-1}u_2, \kappa^{-1}u_3)$$
$$= -c\,(\nabla \times E)((\kappa c)^{-1}u_0, \kappa^{-1}u_1, \kappa^{-1}u_2, \kappa^{-1}u_3) = -\mathcal{E}_0\,\kappa c\,(\nabla \times \psi_{123})(u)\ ,$$
$$0 = (\nabla \cdot E)((\kappa c)^{-1}u_0, \kappa^{-1}u_1, \kappa^{-1}u_2, \kappa^{-1}u_3) = \mathcal{E}_0\,\kappa\,(\nabla \cdot \psi_{123})(u)\ ,$$
$$0 = (\nabla \cdot B)((\kappa c)^{-1}u_0, \kappa^{-1}u_1, \kappa^{-1}u_2, \kappa^{-1}u_3) = \mathcal{E}_0\,\kappa\,(\nabla \cdot \psi_{456})(u)\ .$$

Hence, we obtain an equivalent system to (2.8), given by

$$\frac{\partial \psi_{123}}{\partial u_0} = \nabla \times \psi_{456}\ ,\quad \frac{\partial \psi_{456}}{\partial u_0} = -\nabla \times \psi_{123}\ ,$$
$$\nabla \cdot \psi_{123} = 0\ ,\quad \nabla \cdot \psi_{456} = 0\ , \qquad (2.9)$$

where

$$\psi_{123} := \begin{pmatrix} \psi_1 \\ \psi_2 \\ \psi_3 \end{pmatrix} , \ \psi_{456} := \begin{pmatrix} \psi_4 \\ \psi_5 \\ \psi_6 \end{pmatrix} .$$

The system (2.9) splits into evolution equations, i.e., the two equations in the first row of (2.9) and constraint equations, i.e., the two equations in the second row of (2.9). The evolution equations form a symmetric hyperbolic system

$$\frac{\partial \psi}{\partial t} = -\left(A^1 \cdot \frac{\partial \psi}{\partial u_1} + A^2 \cdot \frac{\partial \psi}{\partial u_2} + A^3 \cdot \frac{\partial \psi}{\partial u_3} \right) ,$$

where the dot denotes matrix multiplication and

$$A^1 := \begin{pmatrix} 0 & 0 & 0 & 0 & 0 & 0 \\ 0 & 0 & 0 & 0 & 0 & 1 \\ 0 & 0 & 0 & 0 & -1 & 0 \\ 0 & 0 & 0 & 0 & 0 & 0 \\ 0 & 0 & -1 & 0 & 0 & 0 \\ 0 & 1 & 0 & 0 & 0 & 0 \end{pmatrix} = i\hat{\hat{\sigma}}_1 ,$$

$$A^2 := \begin{pmatrix} 0 & 0 & 0 & 0 & 0 & -1 \\ 0 & 0 & 0 & 0 & 0 & 0 \\ 0 & 0 & 0 & 1 & 0 & 0 \\ 0 & 0 & 1 & 0 & 0 & 0 \\ 0 & 0 & 0 & 0 & 0 & 0 \\ -1 & 0 & 0 & 0 & 0 & 0 \end{pmatrix} = i\hat{\hat{\sigma}}_2 ,$$

$$A^3 := \begin{pmatrix} 0 & 0 & 0 & 0 & 1 & 0 \\ 0 & 0 & 0 & -1 & 0 & 0 \\ 0 & 0 & 0 & 0 & 0 & 0 \\ 0 & -1 & 0 & 0 & 0 & 0 \\ 1 & 0 & 0 & 0 & 0 & 0 \\ 0 & 0 & 0 & 0 & 0 & 0 \end{pmatrix} = i\hat{\hat{\sigma}}_3 ,$$

where $\hat{\hat{\sigma}}_1, \hat{\hat{\sigma}}_2, \hat{\hat{\sigma}}_3 \in \mathrm{M}(6, \mathbb{C})$ are given by (2.2).

For the analysis of (2.9), we use the methods from the theory of semigroups of operators, specifically, Sect. 7.3 of [4]. For this purpose, we define for every $n \in \mathbb{N}^*$ and every multi-index $\alpha \in \mathbb{N}^n$ the densely-defined linear operator ∂^α in $L^2_{\mathbb{C}}(\mathbb{R}^n)$ by

$$\partial^\alpha := (-1)^{|\alpha|} . \left(C_0^\infty(\mathbb{R}^n, \mathbb{C}) \to L^2_{\mathbb{C}}(\mathbb{R}^n), \ f \mapsto \frac{\partial^\alpha f}{\partial u_\alpha} \right)^* ,$$

where

$$|\alpha| := \sum_{j=1}^{n} \alpha_j .$$

Moreover, we define for $k \in \mathbb{N}$ the Sobolev space

$$W^k_{\mathbb{C}}(\mathbb{R}^n) := \bigcap_{\alpha \in \mathbb{N}^n, |\alpha| \leq k} D(\partial^\alpha) \ .$$

Equipped with the scalar product

$$\langle \, , \rangle_k \ : (W^k_{\mathbb{C}}(\mathbb{R}^n))^2 \to \mathbb{C}$$

defined by

$$\langle f, g \rangle_k := \sum_{\alpha \in \mathbb{N}^n, |\alpha| \leq k} \langle \partial^\alpha f | \partial^\alpha g \rangle_2$$

for all $f, g \in W^k_{\mathbb{C}}(\mathbb{R}^n)$, $W^k_{\mathbb{C}}(\mathbb{R}^n)$ is a Hilbert space. In particular, we denote by $|||\ |||_k$ the norm on $W^k_{\mathbb{C}}(\mathbb{R}^n)$ which is induced by $\langle \, , \rangle_k$. Then, according to Theorem 7.3.1 of [4], we have the following.

Theorem 2.1 *We define the densely-defined linear operator* $A_0 : \left(W^1_{\mathbb{C}}(\mathbb{R}^3)\right)^6 \to X$ *in* $X := \left(L^2_{\mathbb{C}}(\mathbb{R}^3)\right)^6$ *by*

$$A_0\psi := \begin{pmatrix} \sum_{j=1}^6 \left(A^1_{1j}\,\partial^1\psi_j + A^2_{1j}\,\partial^2\psi_j + A^3_{1j}\,\partial^3\psi_j\right) \\ \vdots \\ \sum_{j=1}^6 \left(A^1_{6j}\,\partial^1\psi_j + A^2_{6j}\,\partial^2\psi_j + A^3_{6j}\,\partial^3\psi_j\right) \end{pmatrix} ,$$

$i \in \{1, \dots, 6\}$, *for all* $\psi \in \left(W^1_{\mathbb{C}}(\mathbb{R}^3)\right)^6$. *In addition, we define* $\hat{A}$ *as the trivial operator on* $Z := \left(L^2_{\mathbb{C}}(\mathbb{R}^3)\right)^2$, *i.e.,* $\hat{A}f := 0$ *for all* $f \in Z$. *Finally, we define the dense subspace* Y_0 *of* X *by* $Y_0 := \left(W^1_{\mathbb{C}}(\mathbb{R}^3)\right)^6$, *the* Z*-valued linear operator* $S_0 : Y_0 \to Z$ *in* X *by*

$$S_0\psi := \begin{pmatrix} \partial^1\psi_1 + \partial^2\psi_2 + \partial^3\psi_3 \\ \partial^1\psi_4 + \partial^2\psi_5 + \partial^3\psi_6 \end{pmatrix} ,$$

for all $\psi \in Y_0$ *and the subspace* $\tilde{D}$ *of* $D(A_0) \cap Y_0$ *by*

$$\tilde{D} := \left(W^2_{\mathbb{C}}(\mathbb{R}^3)\right)^6 \ .$$

(i) A_0 *is closable and its closure* A *is skew-symmetric* ($:\Leftrightarrow -iA$ *is self-adjoint*) *and hence the infinitesimal generator of a unitary one-parameter group* $\mathcal{T} : \mathbb{R} \to \mathrm{L}(X)$, *given by* $\mathcal{T}(t) := e^{-it(-iA)}$, *for every* $t \in \mathbb{R}$.
(ii) Define for every $\lambda \in \mathbb{C}^*$ *the bounded linear operator* $B_\lambda : \left(W^2_{\mathbb{C}}(\mathbb{R}^3)\right)^6 \to X$ *by*

$$B_\lambda \psi := \begin{pmatrix} \lambda^{-1}\,\partial^1(\partial^1\psi_1 + \partial^2\psi_2 + \partial^3\psi_3) + \partial^2\psi_6 - \partial^3\psi_5 - \lambda\,\psi_1 \\ \lambda^{-1}\,\partial^2(\partial^1\psi_1 + \partial^2\psi_2 + \partial^3\psi_3) + \partial^3\psi_4 - \partial^1\psi_6 - \lambda\,\psi_2 \\ \lambda^{-1}\,\partial^3(\partial^1\psi_1 + \partial^2\psi_2 + \partial^3\psi_3) + \partial^1\psi_5 - \partial^2\psi_4 - \lambda\,\psi_3 \\ \lambda^{-1}\,\partial^1(\partial^1\psi_4 + \partial^2\psi_5 + \partial^3\psi_6) + \partial^3\psi_2 - \partial^2\psi_3 - \lambda\,\psi_4 \\ \lambda^{-1}\,\partial^2(\partial^1\psi_4 + \partial^2\psi_5 + \partial^3\psi_6) + \partial^1\psi_3 - \partial^3\psi_1 - \lambda\,\psi_5 \\ \lambda^{-1}\,\partial^3(\partial^1\psi_4 + \partial^2\psi_5 + \partial^3\psi_6) + \partial^2\psi_1 - \partial^1\psi_2 - \lambda\,\psi_6 \end{pmatrix},$$

for all $\psi \in \left(W^2_{\mathbb{C}}(\mathbb{R}^3)\right)^6$. *Then*

$$(A - \lambda)^{-1}\psi = B_\lambda \begin{pmatrix} (-\Delta + \lambda^2)^{-1}\psi_1 \\ \vdots \\ (-\Delta + \lambda^2)^{-1}\psi_6 \end{pmatrix},$$

for all $\psi \in X$ *and* $\lambda \in \mathbb{C}\backslash(i\mathbb{R})$ *where* $\Delta : W^2_{\mathbb{C}}(\mathbb{R}^3) \to L^2_{\mathbb{C}}(\mathbb{R}^3)$ *is defined by* $\Delta f := (\partial^1\partial^1 + \partial^2\partial^2 + \partial^3\partial^3) f$ *for all* $f \in W^2_{\mathbb{C}}(\mathbb{R}^3)$.

(iii) S_0 *is closable,*

$$S_0 A_0 \psi = \hat{A} S_0 \psi$$

for all $\psi \in \tilde{D}$ *and for all* $\lambda \in \mathbb{C}\backslash(i\mathbb{R})$

$$(A_0 - \lambda)\tilde{D} = Y_0 \,.$$

(iv) $S := \bar{S}_0$ *satisfies*

$$S\,\mathcal{T}(t) \supset \hat{\mathcal{T}}(t) S$$

where $\hat{\mathcal{T}} = ([0,\infty) \to Z, t \mapsto \mathrm{id}_Z)$ *is the strongly continuous semigroup on* Z *generated by* $\hat{A}$. *In particular,* $\ker S$ *is left invariant by* $\mathcal{T}(t)$ *for every* $t \in [0,\infty)$.

We note that the one-parameter group $\mathcal{T}$ solves the initial value problem of the evolution equations. Since $\mathcal{T}$ leaves the closed subspace $\ker S$ of $(W^1_{\mathbb{C}}(\mathbb{R}^3))^6$ invariant, whose elements solve the constraint equations, the one-parameter group given by the restrictions of $\mathcal{T}(t)$, in domain and in range to $\ker S$, for every $t \in \mathbb{R}$, solve the constrained initial value problem of the evolution equations.

Conclusion 3

From a physics perspective, Part I and Part II of the book construct one-particle theories that provide the basis for the formulation of the quantum field theories of scalar fields, the Dirac field and the electromagnetic field, in Minkowski space, via the method of second quantization. The details of this formulation for the case of the real scalar field of mass $m > 0$, starting from the unitary/anti-unitary representation $\hat{\hat{U}}_a$ from Sect. 5.5 of Part I, are described in [5]. In this construction

$$a = \frac{mc}{\hbar\kappa} = \frac{1}{\kappa\lambda_C} ,$$

where c denotes the speed of light in vacuum, $\hbar$ denotes the reduced Planck's constant, $\kappa > 0$ is a scale factor[1] with dimension 1/length and λ_C denotes the reduced Compton wavelength of the particle. It should be noted that for all these one-particle theories the spectrum of the corresponding Hamilton operators, generating the time evolution in inertial systems, is given by

$$[mc^2, \infty) ,$$

where $m \geqslant 0$ is the rest mass of the particle, as is suitable for a one-particle theory. We note that, as is known, the Hamilton operators, generating the time-evolution of the solutions of the Weyl equations, the Dirac equation and the Maxwell equations are not suitable for this purpose, since not describing single particles. The latter is indicated by the fact that in all these cases the spectrum of the Hamilton operator is not bounded from below, see Sects. 1.7, 1.11

[1] The quantity κ determines the scale in a position representation, see, Sects. 1.1 and 1.3 in [7].

H. R. Beyer, *Quantum Spin and Representations of the Poincaré Group, Part II*,
Synthesis Lectures on Engineering, Science, and Technology,
https://doi.org/10.1007/978-3-031-95823-6_3

and 2.4. It is noteworthy that the generators corresponding to one-parameter groups of Lorentz boosts in the Weyl representations, the Dirac representation and the representation associated with the electromagnetic field are not self-adjoint and hence not observables, whereas they are self-adjoint and hence observables in the representations corresponding to scalar fields. These properties are preserved in the process of second quantization.

Appendix

A.1 Solutions

Solution 1.1 Part (i): According to (1.5), we have that

$$\Phi_2\left(\frac{1}{\sqrt{2}}\begin{pmatrix} e^{i\pi/4} & -e^{-i\pi/4} \\ e^{i\pi/4} & e^{-i\pi/4} \end{pmatrix}\right) = \begin{pmatrix} 1 & 0 & 0 & 0 \\ 0 & 0 & 0 & 1 \\ 0 & 1 & 0 & 0 \\ 0 & 0 & 1 & 0 \end{pmatrix} = U \ .$$

Hence, $U \in \text{Ran}(\Phi_2) = \mathscr{L}_+^{\uparrow}$. Further,

$$U^2 = \begin{pmatrix} 1 & 0 & 0 & 0 \\ 0 & 0 & 0 & 1 \\ 0 & 1 & 0 & 0 \\ 0 & 0 & 1 & 0 \end{pmatrix} \cdot \begin{pmatrix} 1 & 0 & 0 & 0 \\ 0 & 0 & 0 & 1 \\ 0 & 1 & 0 & 0 \\ 0 & 0 & 1 & 0 \end{pmatrix} = \begin{pmatrix} 1 & 0 & 0 & 0 \\ 0 & 0 & 1 & 0 \\ 0 & 0 & 0 & 1 \\ 0 & 1 & 0 & 0 \end{pmatrix} ,$$

$$U^3 = \begin{pmatrix} 1 & 0 & 0 & 0 \\ 0 & 0 & 0 & 1 \\ 0 & 1 & 0 & 0 \\ 0 & 0 & 1 & 0 \end{pmatrix} \cdot \begin{pmatrix} 1 & 0 & 0 & 0 \\ 0 & 0 & 1 & 0 \\ 0 & 0 & 0 & 1 \\ 0 & 1 & 0 & 0 \end{pmatrix} = \begin{pmatrix} 1 & 0 & 0 & 0 \\ 0 & 1 & 0 & 0 \\ 0 & 0 & 1 & 0 \\ 0 & 0 & 0 & 1 \end{pmatrix} = E ,$$

and hence that

$$U^{-1} = U^2 = \begin{pmatrix} 1 & 0 & 0 & 0 \\ 0 & 0 & 1 & 0 \\ 0 & 0 & 0 & 1 \\ 0 & 1 & 0 & 0 \end{pmatrix} .$$

Also, we have that

H. R. Beyer, *Quantum Spin and Representations of the Poincaré Group, Part II*, Synthesis Lectures on Engineering, Science, and Technology,
https://doi.org/10.1007/978-3-031-95823-6

$$U \cdot \begin{pmatrix} \cosh(s) & 0 & 0 & \sinh(s) \\ 0 & 1 & 0 & 0 \\ 0 & 0 & 1 & 0 \\ \sinh(s) & 0 & 0 & \cosh(s) \end{pmatrix} U^{-1}$$

$$= \begin{pmatrix} \cosh(s) & 0 & 0 & \sinh(s) \\ \sinh(s) & 0 & 0 & \cosh(s) \\ 0 & 1 & 0 & 0 \\ 0 & 0 & 1 & 0 \end{pmatrix} \cdot \begin{pmatrix} 1 & 0 & 0 & 0 \\ 0 & 0 & 1 & 0 \\ 0 & 0 & 0 & 1 \\ 0 & 1 & 0 & 0 \end{pmatrix}$$

$$= \begin{pmatrix} \cosh(s) & \sinh(s) & 0 & 0 \\ \sinh(s) & \cosh(s) & 0 & 0 \\ 0 & 0 & 1 & 0 \\ 0 & 0 & 0 & 1 \end{pmatrix},$$

$$U \cdot \begin{pmatrix} \cosh(s) & \sinh(s) & 0 & 0 \\ \sinh(s) & \cosh(s) & 0 & 0 \\ 0 & 0 & 1 & 0 \\ 0 & 0 & 0 & 1 \end{pmatrix} U^{-1}$$

$$= \begin{pmatrix} \cosh(s) & \sinh(s) & 0 & 0 \\ 0 & 0 & 0 & 1 \\ \sinh(s) & \cosh(s) & 0 & 0 \\ 0 & 0 & 1 & 0 \end{pmatrix} \cdot \begin{pmatrix} 1 & 0 & 0 & 0 \\ 0 & 0 & 1 & 0 \\ 0 & 0 & 0 & 1 \\ 0 & 1 & 0 & 0 \end{pmatrix}$$

$$= \begin{pmatrix} \cosh(s) & 0 & \sinh(s) & 0 \\ 0 & 1 & 0 & 0 \\ \sinh(s) & 0 & \cosh(s) & 0 \\ 0 & 0 & 0 & 1 \end{pmatrix},$$

for every $s \in \mathbb{R}$.

Part (ii): With the help of (1.5), it follows that

$$\begin{pmatrix} \cosh(s) & \sinh(s) & 0 & 0 \\ \sinh(s) & \cosh(s) & 0 & 0 \\ 0 & 0 & 1 & 0 \\ 0 & 0 & 0 & 1 \end{pmatrix}$$

$$= \Phi_2\left(\frac{1}{\sqrt{2}} \begin{pmatrix} e^{i\pi/4} & -e^{-i\pi/4} \\ e^{i\pi/4} & e^{-i\pi/4} \end{pmatrix} \cdot \begin{pmatrix} e^{s/2} & 0 \\ 0 & e^{-s/2} \end{pmatrix} \cdot \frac{1}{\sqrt{2}} \begin{pmatrix} e^{-i\pi/4} & e^{-i\pi/4} \\ -e^{i\pi/4} & e^{i\pi/4} \end{pmatrix}\right)$$

$$
\begin{aligned}
&= \Phi_2\left(\frac{1}{2}\begin{pmatrix} e^{i\pi/4}e^{s/2} & -e^{-i\pi/4}e^{-s/2} \\ e^{i\pi/4}e^{s/2} & e^{-i\pi/4}e^{-s/2} \end{pmatrix} \cdot \begin{pmatrix} e^{-i\pi/4} & e^{-i\pi/4} \\ -e^{i\pi/4} & e^{i\pi/4} \end{pmatrix}\right) \\
&= \Phi_2\left(\frac{1}{2}\begin{pmatrix} e^{s/2}+e^{-s/2} & e^{s/2}-e^{-s/2} \\ e^{s/2}-e^{-s/2} & e^{s/2}+e^{-s/2} \end{pmatrix}\right) \\
&= \Phi_2\left(\begin{pmatrix} \cosh(s/2) & \sinh(s/2) \\ \sinh(s/2) & \cosh(s/2) \end{pmatrix}\right)
\end{aligned}
$$

and that

$$
\begin{aligned}
&\begin{pmatrix} \cosh(s) & 0 & \sinh(s) & 0 \\ 0 & 1 & 0 & 0 \\ \sinh(s) & 0 & \cosh(s) & 0 \\ 0 & 0 & 0 & 1 \end{pmatrix} \\
&= \Phi_2\left(\frac{1}{2}\begin{pmatrix} e^{i\pi/4} & -e^{-i\pi/4} \\ e^{i\pi/4} & e^{-i\pi/4} \end{pmatrix} \cdot \begin{pmatrix} \cosh(s/2) & \sinh(s/2) \\ \sinh(s/2) & \cosh(s/2) \end{pmatrix} \cdot \begin{pmatrix} e^{-i\pi/4} & e^{-i\pi/4} \\ -e^{i\pi/4} & e^{i\pi/4} \end{pmatrix}\right) \\
&= \Phi_2\left(\frac{1}{2}\begin{pmatrix} e^{i\pi/4}\cosh(s/2) - e^{-i\pi/4}\sinh(s/2) & e^{i\pi/4}\sinh(s/2) - e^{-i\pi/4}\cosh(s/2) \\ e^{i\pi/4}\cosh(s/2) + e^{-i\pi/4}\sinh(s/2) & e^{i\pi/4}\sinh(s/2) + e^{-i\pi/4}\cosh(s/2) \end{pmatrix} \cdot \begin{pmatrix} e^{-i\pi/4} & e^{-i\pi/4} \\ -e^{i\pi/4} & e^{i\pi/4} \end{pmatrix}\right) \\
&= \Phi_2\left(\frac{1}{2}\begin{pmatrix} \cosh(s/2) - e^{-i\pi/2}\sinh(s/2) - e^{i\pi/2}\sinh(s/2) + \cosh(s/2) & \cosh(s/2) - e^{-i\pi/2}\sinh(s/2) + e^{i\pi/2}\sinh(s/2) - \cosh(s/2) \\ \cosh(s/2) + e^{-i\pi/2}\sinh(s/2) - e^{i\pi/2}\sinh(s/2) - \cosh(s/2) & \cosh(s/2) + e^{-i\pi/2}\sinh(s/2) + e^{i\pi/2}\sinh(s/2) + \cosh(s/2) \end{pmatrix}\right) \\
&= \Phi_2\left(\begin{pmatrix} \cosh(s/2) & i\sinh(s/2) \\ -i\sinh(s/2) & \cosh(s/2) \end{pmatrix}\right) .
\end{aligned}
$$

Solution 1.2 For

$$
G = \begin{pmatrix} \alpha & \beta \\ \gamma & \delta \end{pmatrix} \in \mathrm{SL}(2,\mathbb{C}) ,
$$

according to (1.6), we have that

$$
\begin{aligned}
&\Phi_2(G) \\
&= \begin{pmatrix} \frac{1}{2}(|\alpha|^2+|\beta|^2+|\gamma|^2+|\delta|^2) & \mathrm{Re}(\alpha\beta^*+\gamma\delta^*) & \mathrm{Im}(\alpha\beta^*+\gamma\delta^*) & \frac{1}{2}(|\alpha|^2-|\beta|^2+|\gamma|^2-|\delta|^2) \\ \mathrm{Re}(\alpha\gamma^*+\beta\delta^*) & \mathrm{Re}(\alpha\delta^*+\beta\gamma^*) & \mathrm{Im}(\alpha\delta^*-\beta\gamma^*) & \mathrm{Re}(\alpha\gamma^*-\beta\delta^*) \\ -\mathrm{Im}(\alpha\gamma^*+\beta\delta^*) & -\mathrm{Im}(\alpha\delta^*+\beta\gamma^*) & \mathrm{Re}(\alpha\delta^*-\beta\gamma^*) & -\mathrm{Im}(\alpha\gamma^*-\beta\delta^*) \\ \frac{1}{2}(|\alpha|^2+|\beta|^2-|\gamma|^2-|\delta|^2) & \mathrm{Re}(\alpha\beta^*-\gamma\delta^*) & \mathrm{Im}(\alpha\beta^*-\gamma\delta^*) & \frac{1}{2}(|\alpha|^2-|\beta|^2-|\gamma|^2+|\delta|^2) \end{pmatrix} ,
\end{aligned}
$$

Hence,

$$
\begin{aligned}
&\Phi_2(G^*) \\
&= \begin{pmatrix} \frac{1}{2}(|\alpha^*|^2+|\gamma|^2+|\beta|^2+|(\delta^*)|^2) & \mathrm{Re}(\alpha^*\gamma+\beta^*(\delta^*)^*) & \mathrm{Im}(\alpha^*\gamma+\beta^*(\delta^*)^*) & \frac{1}{2}(|\alpha^*|^2-|\gamma|^2+|\beta|^2-|(\delta^*)|^2) \\ \mathrm{Re}(\alpha^*\beta+\gamma^*(\delta^*)^*) & \mathrm{Re}(\alpha^*(\delta^*)^*+\gamma^*\beta) & \mathrm{Im}(\alpha^*(\delta^*)^*-\gamma^*\beta) & \mathrm{Re}(\alpha^*\beta-\gamma^*(\delta^*)^*) \\ -\mathrm{Im}(\alpha^*\beta+\gamma^*(\delta^*)^*) & -\mathrm{Im}(\alpha^*(\delta^*)^*+\gamma^*\beta) & \mathrm{Re}(\alpha^*(\delta^*)^*-\gamma^*\beta) & -\mathrm{Im}(\alpha^*\beta-\gamma^*(\delta^*)^*) \\ \frac{1}{2}(|\alpha^*|^2+|\gamma|^2-|\beta|^2-|(\delta^*)|^2) & \mathrm{Re}(\alpha^*\gamma-\beta^*(\delta^*)^*) & \mathrm{Im}(\alpha^*\gamma-\beta^*(\delta^*)^*) & \frac{1}{2}(|\alpha^*|^2-|\gamma|^2-|\beta|^2+|(\delta^*)|^2) \end{pmatrix} \\
&= \begin{pmatrix} \frac{1}{2}(|\alpha|^2+|\beta|^2+|\gamma|^2+|\delta|^2) & \mathrm{Re}(\alpha^*\gamma+\beta^*\delta) & \mathrm{Im}(\alpha^*\gamma+\beta^*\delta) & \frac{1}{2}(|\alpha|^2+|\beta|^2-|\gamma|^2-|\delta|^2) \\ \mathrm{Re}(\alpha^*\beta+\gamma^*\delta) & \mathrm{Re}(\alpha^*\delta+\gamma^*\beta) & \mathrm{Im}(\alpha^*\delta-\gamma^*\beta) & \mathrm{Re}(\alpha^*\beta-\gamma^*\delta) \\ -\mathrm{Im}(\alpha^*\beta+\gamma^*\delta) & -\mathrm{Im}(\alpha^*\delta+\gamma^*\beta) & \mathrm{Re}(\alpha^*\delta-\gamma^*\beta) & -\mathrm{Im}(\alpha^*\beta-\gamma^*\delta) \\ \frac{1}{2}(|\alpha|^2-|\beta|^2+|\gamma|^2-|\delta|^2) & \mathrm{Re}(\alpha^*\gamma-\beta^*\delta) & \mathrm{Im}(\alpha^*\gamma-\beta^*\delta) & \frac{1}{2}(|\alpha|^2-|\beta|^2-|\gamma|^2+|\delta|^2) \end{pmatrix} ,
\end{aligned}
$$

and

$$
\begin{aligned}
&[\Phi_2(G^*)]^t \\
&= \begin{pmatrix}
\frac{1}{2}\left(|\alpha|^2+|\beta|^2+|\gamma|^2+|\delta|^2\right) & \mathrm{Re}(\alpha^*\beta+\gamma^*\delta) & -\mathrm{Im}(\alpha^*\beta+\gamma^*\delta) & \frac{1}{2}\left(|\alpha|^2-|\beta|^2+|\gamma|^2-|\delta|^2\right) \\
\mathrm{Re}(\alpha^*\gamma+\beta^*\delta) & \mathrm{Re}(\alpha^*\delta+\gamma^*\beta) & -\mathrm{Im}(\alpha^*\delta+\gamma^*\beta) & \mathrm{Re}(\alpha^*\gamma-\beta^*\delta) \\
\mathrm{Im}(\alpha^*\gamma+\beta^*\delta) & \mathrm{Im}(\alpha^*\delta-\gamma^*\beta) & \mathrm{Re}(\alpha^*\delta-\gamma^*\beta) & \mathrm{Im}(\alpha^*\gamma-\beta^*\delta) \\
\frac{1}{2}\left(|\alpha|^2+|\beta|^2-|\gamma|^2-|\delta|^2\right) & \mathrm{Re}(\alpha^*\beta-\gamma^*\delta) & -\mathrm{Im}(\alpha^*\beta-\gamma^*\delta) & \frac{1}{2}\left(|\alpha|^2-|\beta|^2-|\gamma|^2+|\delta|^2\right)
\end{pmatrix} \\
&= \begin{pmatrix}
\frac{1}{2}\left(|\alpha|^2+|\beta|^2+|\gamma|^2+|\delta|^2\right) & \mathrm{Re}(\alpha\beta^*+\gamma\delta^*) & \mathrm{Im}(\alpha\beta^*+\gamma\delta^*) & \frac{1}{2}\left(|\alpha|^2-|\beta|^2+|\gamma|^2-|\delta|^2\right) \\
\mathrm{Re}(\alpha\gamma^*+\beta\delta^*) & \mathrm{Re}(\alpha\delta^*+\gamma\beta^*) & \mathrm{Im}(\alpha\delta^*+\gamma\beta^*) & \mathrm{Re}(\alpha\gamma^*-\beta\delta^*) \\
-\mathrm{Im}(\alpha\gamma^*+\beta\delta^*) & -\mathrm{Im}(\alpha\delta^*-\gamma\beta^*) & \mathrm{Re}(\alpha\delta^*-\gamma\beta^*) & -\mathrm{Im}(\alpha\gamma^*-\beta\delta^*) \\
\frac{1}{2}\left(|\alpha|^2+|\beta|^2-|\gamma|^2-|\delta|^2\right) & \mathrm{Re}(\alpha\beta^*-\gamma\delta^*) & \mathrm{Im}(\alpha\beta^*-\gamma\delta^*) & \frac{1}{2}\left(|\alpha|^2-|\beta|^2-|\gamma|^2+|\delta|^2\right)
\end{pmatrix} .
\end{aligned}
$$

Since,

$$
\begin{aligned}
\mathrm{Re}(\alpha\delta^*+\gamma\beta^*) &= \mathrm{Re}(\alpha\delta^*+\beta\gamma^*)\ ,\ \ \mathrm{Im}(\alpha\delta^*+\gamma\beta^*) = \mathrm{Im}(\alpha\delta^*-\beta\gamma^*)\ , \\
-\mathrm{Im}(\alpha\delta^*-\gamma\beta^*) &= -\mathrm{Im}(\alpha\delta^*+\beta\gamma^*)\ ,\ \ \mathrm{Re}(\alpha\delta^*-\gamma\beta^*) = \mathrm{Re}(\alpha\delta^*-\beta\gamma^*)\ ,
\end{aligned}
$$

it follows that

$$[\Phi_2(G^*)]^t = \Phi_2(G)$$

and hence that

$$\Phi_2(G^*) = [\Phi_2(G)]^t\ .$$

Further, according to (1.11), for $l \in \{0, 1, 2, 3\}$, we have that

$$\sum_{k=0}^{3} [\Phi_2(G)]_{kl}\, \sigma_k = G \cdot \sigma_l \cdot G^*\ .$$

Hence,

$$\sum_{k=0}^{3} [\Phi_2(G)]_{lk}\, \sigma_k = \sum_{k=0}^{3} [[\Phi_2(G)]^t]_{kl}\, \sigma_k = \sum_{k=0}^{3} [\Phi_2(G^*)]_{kl}\, \sigma_k = G^* \cdot \sigma_l \cdot G\ .$$

Solution 1.3 Part (i): For every

$$G = \begin{pmatrix} \alpha & \beta \\ \gamma & \delta \end{pmatrix} \in \mathrm{SL}(2, \mathbb{C})\ ,$$

we have that

$$G_{cc} = \begin{pmatrix} \alpha^* & \beta^* \\ \gamma^* & \delta^* \end{pmatrix} ,$$

$$(-G)_{cc} = \begin{pmatrix} -\alpha & -\beta \\ -\gamma & -\delta \end{pmatrix}_{cc} = \begin{pmatrix} -\alpha^* & -\beta^* \\ -\gamma^* & -\delta^* \end{pmatrix} = -\begin{pmatrix} \alpha^* & \beta^* \\ \gamma^* & \delta^* \end{pmatrix} = -G_{cc}$$

and

$$\det(G_{cc}) = \begin{vmatrix} \alpha^* & \beta^* \\ \gamma^* & \delta^* \end{vmatrix} = \alpha^*\delta^* - \beta^*\gamma^* = (\alpha\delta - \beta\gamma)^* = (\det(G))^* = 1 .$$

Therefore, $\hbar_{cc}$ is well-defined, $\hbar_{cc}(-G) = -\hbar_{cc}(G)$, for every $G \in \mathrm{SL}(2, \mathbb{C})$, and it follows that $\hbar_{cc}$ is component-wise continuous. Further, since

$$(G_1 \cdot G_2)_{cc} = ((G_1 \cdot G_2)^t)^* = (G_2^t \cdot G_1^t)^* = (G_1^t)^* \cdot (G_2^t)^* = G_{1cc} \cdot G_{2cc} ,$$

for all $G_1, G_2 \in \mathrm{SL}(2, \mathbb{C})$, it follows that $\hbar_{cc}$ is a group homomorphism. Also, since $(G_{cc})_{cc} = G$, for every $G \in \mathrm{SL}(2, \mathbb{C})$, $\hbar_{cc}$ is an involutory group isomorphism. Finally, if $G \in \mathrm{SU}(2, \mathbb{C})$, then

$$G = \begin{pmatrix} U_{11} & U_{12} \\ -U_{12}^* & U_{11}^* \end{pmatrix} ,$$

where $U_{11}, U_{12} \in \mathbb{C}$ such that $|U_{11}|^2 + |U_{12}|^2 = 1$. Hence,

$$G_{cc} = \begin{pmatrix} U_{11}^* & U_{12}^* \\ -(U_{12}^*)^* & (U_{11}^*)^* \end{pmatrix} \in \mathrm{SU}(2, \mathbb{C}) .$$

Part (ii): For every

$$G = \begin{pmatrix} \alpha & \beta \\ \gamma & \delta \end{pmatrix} \in \mathrm{SL}(2, \mathbb{C}) ,$$

we have that

$$\det(G^*) = \begin{vmatrix} \alpha^* & \gamma^* \\ \beta^* & \delta^* \end{vmatrix} = \alpha^*\delta^* - \gamma^*\beta^* = (\alpha\delta - \beta\gamma)^* = (\det(G))^* = 1$$

and hence that

$$(G^*)^{-1} = \begin{pmatrix} \delta^* & -\gamma^* \\ -\beta^* & \alpha^* \end{pmatrix} \in \mathrm{SL}(2, \mathbb{C}) .$$

Also,

$$[(-G)^*]^{-1} = -(G^*)^{-1} .$$

Therefore, $\hbar_{*^{-1}}$ is well-defined, $\hbar_{*^{-1}}(-G) = -\hbar_{*^{-1}}(G)$, for every $G \in \mathrm{SL}(2, \mathbb{C})$, and it follows that $\hbar_{*^{-1}}$ is component-wise continuous. Since,

$$\begin{aligned} \hbar_{*^{-1}}(G_1 \cdot G_2) &:= ((G_1 \cdot G_2)^*)^{-1} = (G_2^* \cdot G_1^*)^{-1} \\ &= (G_1^*)^{-1} \cdot (G_2^*)^{-1} = \hbar_{*^{-1}}(G_1) \cdot \hbar_{*^{-1}}(G_2) , \end{aligned}$$

for all $G_1, G_2 \in \mathrm{SL}(2,\mathbb{C})$, $\hbar_{*^{-1}}$ is group homomorphism. Further, since

$$\hbar_{*^{-1}}((G^{-1})^*) = (((G^{-1})^*)^*)^{-1} = G \ ,$$

for every $G \in \mathrm{SL}(2,\mathbb{C})$ and since, if $G_1, G_2 \in \mathrm{SL}(2,\mathbb{C})$ are such that $\hbar_{*^{-1}}(G_1) = \hbar_{*^{-1}}(G_2)$, then

$$(G_1^*)^{-1} = (G_2^*)^{-1} \ ,$$

and therefore $G_1^* = G_2^*$ and $G_1 = G_2$, $\hbar_{*^{-1}}$ is a group automorphism. Finally, if $G \in \mathrm{SU}(2,\mathbb{C})$, then

$$G = \begin{pmatrix} U_{11} & U_{12} \\ -U_{12}^* & U_{11}^* \end{pmatrix} \ ,$$

where $U_{11}, U_{12} \in \mathbb{C}$ are such that $|U_{11}|^2 + |U_{12}|^2 = 1$, and hence

$$G^* = \begin{pmatrix} U_{11}^* & -U_{12} \\ U_{12}^* & U_{11} \end{pmatrix} \ , \ (G^*)^{-1} = \begin{pmatrix} U_{11} & U_{12} \\ -U_{12}^* & U_{11}^* \end{pmatrix} = G \ .$$

Finally, for

$$G = \begin{pmatrix} \alpha & \beta \\ \gamma & \delta \end{pmatrix} \in \mathrm{SL}(2,\mathbb{C}) \ ,$$

we have that

$$G^{-1} = \begin{pmatrix} \delta & -\beta \\ -\gamma & \alpha \end{pmatrix} \ , \ (G^{-1})^* = \begin{pmatrix} \delta^* & -\gamma^* \\ -\beta^* & \alpha^* \end{pmatrix} = (G^*)^{-1} \ .$$

Solution 1.4 Since $\sigma_1^2 = \sigma_2^2 = \sigma_3^2 = E$, for $s \in \mathbb{C}$, we have that

$$\begin{aligned} \exp(s\sigma_j) &= \sum_{k=0}^{\infty} \frac{s^k}{k!}\sigma_j^k = \sum_{k=0}^{\infty} \frac{s^{2k}}{(2k)!}\sigma_j^{2k} + \sum_{k=0}^{\infty} \frac{s^{2k+1}}{(2k+1)!}\sigma_j^{2k+1} \\ &= \left[\sum_{k=0}^{\infty} \frac{s^{2k}}{(2k)!}\right] E + \left[\sum_{k=0}^{\infty} \frac{s^{2k+1}}{(2k+1)!}\right] \sigma_j = \cosh(s).E + \sinh(s).\sigma_j \ . \end{aligned}$$

Hence,

$$\exp(is\sigma_1) = \cos(s).E + i\sin(s).\sigma_1 = \begin{pmatrix} \cos(s) & i\sin(s) \\ i\sin(s) & \cos(s) \end{pmatrix} \ ,$$

$$\exp(is\sigma_2) = \cos(s).E + i\sin(s).\sigma_2 = \begin{pmatrix} \cos(s) & \sin(s) \\ -\sin(s) & \cos(s) \end{pmatrix} \ ,$$

$$\begin{aligned} \exp(is\sigma_3) &= \cos(s).E + i\sin(s).\sigma_3 = \cos(s).E + i\sin(s)\begin{pmatrix} 1 & 0 \\ 0 & -1 \end{pmatrix} \\ &= \begin{pmatrix} e^{is} & 0 \\ 0 & e^{-is} \end{pmatrix} \ , \end{aligned}$$

$$\exp(s\sigma_1) = \cosh(s).E + \sinh(s).\sigma_1 = \begin{pmatrix} \cosh(s) & \sinh(s) \\ \sinh(s) & \cosh(s) \end{pmatrix} ,$$

$$\exp(s\sigma_2) = \cosh(s).E + \sinh(s).\sigma_2 = \begin{pmatrix} \cosh(s) & -i\sinh(s) \\ i\sinh(s) & \cosh(s) \end{pmatrix} ,$$

$$\begin{aligned} \exp(s\sigma_3) &= \cosh(s).E + \sinh(s).\sigma_3 \\ &= \begin{pmatrix} \cosh(s) + \sinh(s) & 0 \\ 0 & \cosh(s) - \sinh(s) \end{pmatrix} = \begin{pmatrix} e^s & 0 \\ 0 & e^{-s} \end{pmatrix} . \end{aligned}$$

Solution 1.5 According to Part I, for $a \geqslant 0$, we have that

$$\mathcal{V}_a \hat{L}_3 \mathcal{V}_a^{-1} = \hat{\mathfrak{L}}_3 ,$$

where $\mathcal{V}_a : L^2_{\mathbb{C}}(\mathbb{R}^3, \varphi_a) \to L^2_{\mathbb{C}}(\mathbb{R}^3)$ is the Hilbert space isomorphism from Exercise 5.17 in Part I and $\hat{\mathfrak{L}}_3$ is the 3rd component of angular momentum from quantum mechanics, e.g., see Sect. 2.4 in [7]. Hence, $\mathcal{V}_a \times \mathcal{V}_a : (L^2_{\mathbb{C}}(\mathbb{R}^3, \varphi_a))^2 \to (L^2_{\mathbb{C}}(\mathbb{R}^3))^2$ is a Hilbert space isomorphism and, for $j \in \{1, 2, 3\}$, we have that

$$(\mathcal{V}_a \times \mathcal{V}_a)\hat{J}_j(\mathcal{V}_a \times \mathcal{V}_a)^{-1} = \hat{\mathfrak{J}}_j ,$$

where $\hat{\mathfrak{J}}_j : D(\hat{\mathfrak{L}}_j) \times D(\hat{\mathfrak{L}}_j) \to (L^2_{\mathbb{C}}(\mathbb{R}^3))^2$ is defined by

$$\hat{\mathfrak{J}}_j f := \begin{pmatrix} \hat{\mathfrak{L}}_j f_1 \\ \hat{\mathfrak{L}}_j f_2 \end{pmatrix} + \frac{\hbar}{2}\sigma_j \cdot \begin{pmatrix} f_1 \\ f_2 \end{pmatrix} ,$$

for every $f = {}^t(f_1, f_2) \in D(\hat{\mathfrak{L}}_j) \times D(\hat{\mathfrak{L}}_j)$. Further, in [7], the introduction of spherical coordinates induces an unitary transformation

$$U : L^2_{\mathbb{C}}(\mathbb{R}^3) \to X := L^2_{\mathbb{C}}(\Omega, u^2 \sin(\theta)) ,$$

where $\Omega := (0, \infty) \times (0, \pi) \times (-\pi, \pi)$, and u and θ denote the coordinate projections onto the first and second coordinate, respectively. Then, X is decomposed into pairwise orthogonal complex Hilbert spaces $X_{\ell m}$, $(\ell, m) \in \mathcal{I}$, where the index set $\mathcal{I}$ is defined by

$$\mathcal{I} := \bigcup_{\bar{\ell} \in \mathbb{N}} \{(\bar{\ell}, \bar{m}) : \bar{m} \in \{-\bar{\ell}, -\bar{\ell} + 1, \dots, \bar{\ell} - 1, \bar{\ell}\}\} ,$$

given by the range of the linear isometry $U_{\ell m} : L^2_{\mathbb{C}}(I, u^2) \to X$, defined by

$$U_{\ell m} f := f \otimes Y_{\ell m} ,$$

for all $f \in L^2_{\mathbb{C}}(I, u^2)$, where $Y_{\ell m}$ is a spherical harmonic. In addition, for every $(\ell, m) \in \mathcal{I}$ and $f \in L^2_{\mathbb{C}}(I, u^2)$, it is shown that $f \otimes Y_{\ell m}$ is part of the domain of $U \hat{\mathfrak{L}}_3 U^{-1}$ as well as that

$$U\hat{\mathfrak{L}}_3\,U^{-1}f\otimes Y_{\ell m}=m\hbar f\otimes Y_{\ell m}\ .$$

Hence, it follows that $U\times U:(L^2_{\mathbb{C}}(\mathbb{R}^3))^2\to X^2$ is a Hilbert space isomorphism, that

$$\begin{aligned}
&(U\times U)(\hat{\mathfrak{J}}_3-\lambda_1)(U\times U)^{-1}\begin{pmatrix}f\otimes Y_{\ell m}\\0\end{pmatrix}\\
&=(U\times U)\hat{\mathfrak{J}}_3(U\times U)^{-1}\begin{pmatrix}f\otimes Y_{\ell m}\\0\end{pmatrix}-\lambda_1\begin{pmatrix}f\otimes Y_{\ell m}\\0\end{pmatrix}\\
&=(U\times U)\hat{\mathfrak{J}}_3\begin{pmatrix}U^{-1}(f\otimes Y_{\ell m})\\0\end{pmatrix}-\lambda_1\begin{pmatrix}f\otimes Y_{\ell m}\\0\end{pmatrix}\\
&=(U\times U)\left[\begin{pmatrix}\hat{\mathfrak{L}}_3U^{-1}(f\otimes Y_{\ell m})\\0\end{pmatrix}+\frac{\hbar}{2}\sigma_3\cdot\begin{pmatrix}U^{-1}(f\otimes Y_{\ell m})\\0\end{pmatrix}\right]\\
&\quad-\lambda_1\begin{pmatrix}f\otimes Y_{\ell m}\\0\end{pmatrix}\\
&=(U\times U)\left[\begin{pmatrix}\hat{\mathfrak{L}}_3U^{-1}(f\otimes Y_{\ell m})\\0\end{pmatrix}+\frac{\hbar}{2}\begin{pmatrix}U^{-1}(f\otimes Y_{\ell m})\\0\end{pmatrix}\right]-\lambda_1\begin{pmatrix}f\otimes Y_{\ell m}\\0\end{pmatrix}\\
&=\begin{pmatrix}U\hat{\mathfrak{L}}_3U^{-1}(f\otimes Y_{\ell m})\\0\end{pmatrix}+\frac{\hbar}{2}\begin{pmatrix}f\otimes Y_{\ell m}\\0\end{pmatrix}-\lambda_1\begin{pmatrix}f\otimes Y_{\ell m}\\0\end{pmatrix}\\
&=\left[m\hbar+\frac{\hbar}{2}-\hbar\left(m+\frac{1}{2}\right)\right]\begin{pmatrix}f\otimes Y_{\ell m}\\0\end{pmatrix}=0\,,
\end{aligned}$$

and that

$$\begin{aligned}
&(U\times U)(\hat{\mathfrak{J}}_3-\lambda_2)(U\times U)^{-1}\begin{pmatrix}0\\f\otimes Y_{\ell m}\end{pmatrix}\\
&=(U\times U)\hat{\mathfrak{J}}_3(U\times U)^{-1}\begin{pmatrix}0\\f\otimes Y_{\ell m}\end{pmatrix}-\lambda_2\begin{pmatrix}0\\f\otimes Y_{\ell m}\end{pmatrix}\\
&=(U\times U)\hat{\mathfrak{J}}_3\begin{pmatrix}0\\U^{-1}(f\otimes Y_{\ell m})\end{pmatrix}-\lambda_2\begin{pmatrix}0\\f\otimes Y_{\ell m}\end{pmatrix}\\
&=(U\times U)\left[\begin{pmatrix}0\\\hat{\mathfrak{L}}_3U^{-1}(f\otimes Y_{\ell m})\end{pmatrix}+\frac{\hbar}{2}\sigma_3\cdot\begin{pmatrix}0\\U^{-1}(f\otimes Y_{\ell m})\end{pmatrix}\right]\\
&\quad-\lambda_2\begin{pmatrix}0\\f\otimes Y_{\ell m}\end{pmatrix}\\
&=(U\times U)\left[\begin{pmatrix}0\\\hat{\mathfrak{L}}_3U^{-1}(f\otimes Y_{\ell m})\end{pmatrix}-\frac{\hbar}{2}\begin{pmatrix}0\\U^{-1}(f\otimes Y_{\ell m})\end{pmatrix}\right]-\lambda_2\begin{pmatrix}0\\f\otimes Y_{\ell m}\end{pmatrix}
\end{aligned}$$

$$= \begin{pmatrix} 0 \\ U\hat{\mathfrak{L}}_3 U^{-1}(f \otimes Y_{\ell m}) \end{pmatrix} - \frac{\hbar}{2}\begin{pmatrix} 0 \\ f \otimes Y_{\ell m} \end{pmatrix} - \lambda_2 \begin{pmatrix} 0 \\ f \otimes Y_{\ell m} \end{pmatrix}$$

$$= \left[m\hbar - \frac{\hbar}{2} - \hbar\left(m - \frac{1}{2}\right)\right]\begin{pmatrix} 0 \\ f \otimes Y_{\ell m} \end{pmatrix} = 0 \,,$$

where

$$\lambda_1 := \hbar\left(m + \frac{1}{2}\right) \,, \quad \lambda_2 = \hbar\left(m - \frac{1}{2}\right) .$$

Hence, it follows for every $(\ell, m) \in \mathcal{I}$ that the set

$$\bigcup_{(\ell,m)\in\mathcal{I}} \left\{\hbar\left(m + \frac{1}{2}\right), \hbar\left(m - \frac{1}{2}\right)\right\}$$

consist of eigenvalues of $\hat{\mathfrak{J}}_3$ of infinite multiplicity and that there is a Hilbert basis of $(L^2_{\mathbb{C}}(\mathbb{R}^3))^2$ consisting of corresponding eigenvectors. Hence, the spectrum of $\hat{\mathfrak{J}}_3$ is a pure point spectrum, consisting of eigenvalues of infinite multiplicity, given by

$$\hbar\left(\mathbb{Z} + \frac{1}{2}\right) .$$

Solution 1.6 For $\lambda \in \mathbb{C}$, we have that

$$\left(\hat{J}_{R03} - \lambda\right) f = \begin{pmatrix} \hat{L}_{03} f_1 \\ \hat{L}_{03} f_2 \end{pmatrix} - \frac{\hbar}{2} i\sigma_3 \cdot \begin{pmatrix} f_1 \\ f_2 \end{pmatrix} - \lambda \begin{pmatrix} f_1 \\ f_2 \end{pmatrix} = \begin{pmatrix} (\hat{L}_{03} - \frac{\hbar}{2} i - \lambda) f_1 \\ (\hat{L}_{03} + \frac{\hbar}{2} i - \lambda) f_2 \end{pmatrix} ,$$

$$\left(\hat{J}_{L03} - \lambda\right) f = \begin{pmatrix} \hat{L}_{03} f_1 \\ \hat{L}_{03} f_2 \end{pmatrix} + \frac{\hbar}{2} i\sigma_3 \cdot \begin{pmatrix} f_1 \\ f_2 \end{pmatrix} - \lambda \begin{pmatrix} f_1 \\ f_2 \end{pmatrix} = \begin{pmatrix} (\hat{L}_{03} + \frac{\hbar}{2} i - \lambda) f_1 \\ (\hat{L}_{03} - \frac{\hbar}{2} i - \lambda) f_2 \end{pmatrix} ,$$

for every $f = {}^t(f_1, f_2) \in D(\hat{L}_{03}) \times D(\hat{L}_{03})$. Since, according to Part I, the spectrum of $\hat{L}_{03}$ is given by $\mathbb{R}$, it follows that λ is part of the spectrum of $\hat{J}_{R03}$ if and only if

$$\lambda \in \hbar\left(\frac{i}{2} + \mathbb{R}\right) \cup \hbar\left(-\frac{i}{2} + \mathbb{R}\right) .$$

The same reasoning applies to the spectrum of $\hat{J}_{L03}$.

Solution 1.7 If $(a_1, G_1), (a_2, G_2), \ldots$ is a sequence in $\mathbb{R}^4 \times \mathrm{SL}(2, \mathbb{C})$ that is component-wise convergent to $(a, G) \in \mathbb{R}^4 \times \mathrm{M}(2, \mathbb{C})$, it follows that the sequence $G_1, G_2, \ldots$ is component-wise convergent to G and hence, since $\mathrm{SL}(2, \mathbb{C})$ is a closed subset of $\mathrm{M}(2, \mathbb{C})$, that $G \in \mathrm{SL}(2, \mathbb{C})$. Hence, $(a, G) \in \mathbb{R}^4 \times \mathrm{SL}(2, \mathbb{C})$. Further, for every $\nu \in \mathbb{N}$, we have that $({}^t(\nu, 0, 0, 0), E) \in \mathbb{R}^4 \times \mathrm{SL}(2, \mathbb{C})$ and that $\|({}^t(\nu, 0, 0, 0), E)\| = (1 + \nu^2)^{1/2} \geqslant \nu$. Hence, $\mathbb{R}^4 \times \mathrm{SL}(2, \mathbb{C})$ is unbounded.

Solution 1.8 First, we note that by (1.22), for every $(a, G) \in \mathbb{R}^4 \times \mathrm{SL}(2, \mathbb{C})$, there is defined a map $\hat{\Phi}_2 : \mathbb{R}^4 \times \mathrm{SL}(2, \mathbb{C}) \to \mathscr{P}_+^\uparrow$. Further, for $(a_1, G_1), (a_2, G_2) \in \mathbb{R}^4 \times \mathrm{SL}(2, \mathbb{C})$, we have that

$$\begin{aligned}
&\hat{\Phi}_2((a_1, G_1) \cdot (a_2, G_2)) = \hat{\Phi}_2(a_1 + \Phi_2(G_1)a_2, G_1 \cdot G_2) \\
&= (a_1 + \Phi_2(G_1)a_2, \Phi_2(G_1 \cdot G_2)) = (a_1 + \Phi_2(G_1)a_2, \Phi_2(G_1) \cdot \Phi_2(G_2)) \ , \\
&\hat{\Phi}_2(a_1, G_1) \cdot \hat{\Phi}_2(a_2, G_2) = (a_1, \Phi_2(G_1)) \cdot (a_2, \Phi_2(G_2)) \\
&= (a_1 + \Phi_2(G_1)a_2, \Phi_2(G_1) \cdot \Phi_2(G_2))
\end{aligned}$$

and hence that

$$\hat{\Phi}_2((a_1, G_1) \cdot (a_2, G_2)) = \hat{\Phi}_2(a_1, G_1) \cdot \hat{\Phi}_2(a_2, G_2) \ .$$

Hence, $\hat{\Phi}_2$ is a homomorphism. Also, if

$$\hat{\Phi}_2(a_1, G_1) = \hat{\Phi}_2(a_2, G_2) \ ,$$

for $(a_1, G_1), (a_2, G_2) \in \mathbb{R}^4 \times \mathrm{SL}(2, \mathbb{C})$, then

$$(a_1, \Phi_2(G_1)) = (a_2, \Phi_2(G_2)) \ .$$

Hence, $a_1 = a_2$ and $\Phi_2(G_1) = \Phi_2(G_2)$. The latter implies that $G_2 = \pm G_1$. Further, if $a_1 = a_2$ and $G_2 = \pm G_1$, then

$$\hat{\Phi}_2(a_1, G_1) = \hat{\Phi}_2(a_2, \pm G_2) = (a_2, \Phi_2(\pm G_2)) = (a_2, \Phi_2(G_2)) = \hat{\Phi}_2(a_2, G_2) \ .$$

Finally, since $\Phi_2 : \mathrm{SL}(2, \mathbb{C}) \to \mathscr{L}_+^\uparrow$ is component-wise continuous, it also follows that $\hat{\Phi}_2$ is component-wise continuous.

Solution 1.9 For $\lambda \in \mathbb{C}$, we have that

$$\begin{aligned}
&\left((T_{a_0 v_a^0 - \vec{a} \cdot \mathrm{id}_{\mathbb{R}^3}} \times T_{a_0 v_a^0 - \vec{a} \cdot \mathrm{id}_{\mathbb{R}^3}}) - \lambda\right) f = \begin{pmatrix} T_{a_0 v_a^0 - \vec{a} \cdot \mathrm{id}_{\mathbb{R}^3}} f_1 \\ T_{a_0 v_a^0 - \vec{a} \cdot \mathrm{id}_{\mathbb{R}^3}} f_2 \end{pmatrix} - \lambda \begin{pmatrix} f_1 \\ f_2 \end{pmatrix} \\
&= \begin{pmatrix} (T_{a_0 v_a^0 - \vec{a} \cdot \mathrm{id}_{\mathbb{R}^3}} - \lambda) f_1 \\ (T_{a_0 v_a^0 - \vec{a} \cdot \mathrm{id}_{\mathbb{R}^3}} - \lambda) f_2 \end{pmatrix} ,
\end{aligned}$$

for every $f = {}^t(f_1, f_2) \in D(T_{a_0 v_a^0 - \vec{a} \cdot \mathrm{id}_{\mathbb{R}^3}}) \times D(T_{a_0 v_a^0 - \vec{a} \cdot \mathrm{id}_{\mathbb{R}^3}})$. Hence the operator $(T_{a_0 v_a^0 - \vec{a} \cdot \mathrm{id}_{\mathbb{R}^3}} \times T_{a_0 v_a^0 - \vec{a} \cdot \mathrm{id}_{\mathbb{R}^3}}) - \lambda$ is bijective if and only if λ is part of the resolvent set of $T_{a_0 v_a^0 - \vec{a} \cdot \mathrm{id}_{\mathbb{R}^3}}$. Since the spectrum of the latter operator is given by the closure of the range of $a_0 v_a^0 - \vec{a} \cdot \mathrm{id}_{\mathbb{R}^3}$, it follows that λ is part of the spectrum of $T_{a_0 v_a^0 - \vec{a} \cdot \mathrm{id}_{\mathbb{R}^3}} \times T_{a_0 v_a^0 - \vec{a} \cdot \mathrm{id}_{\mathbb{R}^3}}$ if and only if λ is part of the closure of the range of $a_0 v_a^0 - \vec{a} \cdot \mathrm{id}_{\mathbb{R}^3}$.

Solution 1.10 If $a \in \mathbb{R}^4$ is future-oriented, timelike, and, in particular, such that $a \cdot a = 1$, it follows from Corollary 5.5 of Part I the existence of a Lorentz transformation $\Lambda \in \mathcal{L}_+^\uparrow$ such that

$$a = \Lambda\,^t(1, 0, 0, 0)\ .$$

Hence, it follows for $v \in \mathbb{R}^3$ that

$$\begin{aligned} a \cdot {}^t(v_a^0(v), v_1, v_2, v_3) &= (\Lambda\,^t(1, 0, 0, 0)) \cdot p_a(v) \\ &= {}^t(1, 0, 0, 0)) \cdot (\Lambda^{-1} p_a(v)) = (\Lambda^{-1} p_a(v))_0 \\ &= (p_a(h_{a\Lambda}(v)))_0 = v_a^0(h_{a\Lambda}(v))\ , \end{aligned}$$

where the parametrization $p_a : \mathbb{R}^3 \to H_{a+}$ and the bijection $h_{a\Lambda} : \mathbb{R}^3 \to \mathbb{R}^3$ are defined in Part I. Sine $h_{a\Lambda}$ is a bijecion, we have that

$$\text{Ran}(a_0 v_a^0 - \vec{a} \cdot \text{id}_{\mathbb{R}^3}) = \text{Ran}(v_a^0)$$

and hence that

$$\overline{\text{Ran}(a_0 v_a^0 - \vec{a} \cdot \text{id}_{\mathbb{R}^3})} = [a, \infty)\ .$$

Solution 1.11 For $\lambda \in \mathbb{C}$, we have that

$$\begin{aligned} &(\hat{J}_3 - \lambda) f \\ &= {}^t\Big(\hat{L}_3 f_1 + \frac{\hbar}{2}[\sigma_3 \cdot {}^t(f_1, f_2)]_1 - \lambda f_1, \hat{L}_3 f_2 + \frac{\hbar}{2}[\sigma_3 \cdot {}^t(f_1, f_2)]_2 - \lambda f_2, \\ &\qquad \hat{L}_3 f_3 + \frac{\hbar}{2}[\sigma_3 \cdot {}^t(f_3, f_4)]_1 - \lambda f_3, \hat{L}_3 f_4 + \frac{\hbar}{2}[\sigma_3 \cdot {}^t(f_3, f_4)]_2 - \lambda f_4\Big) \\ &= {}^t([(\hat{J}_{W3} - \lambda)\,^t(f_1, f_2)]_1, [(\hat{J}_{W3} - \lambda)\,^t(f_1, f_2)]_2, \\ &\qquad [(\hat{J}_{W3} - \lambda)\,^t(f_3, f_4)]_1, [(\hat{J}_{W3} - \lambda)\,^t(f_3, f_4)]_2) \end{aligned}$$

for every $f \in (D(\hat{L}_3))^4$, where $(\hat{J}_W)_3$ is the 3rd component of total angular momentum defined by (1.18), as a generator of rotations. As a consequence, $\hat{J}_3 - \lambda$ is bijective if and only if $(\hat{J}_W)_3 - \lambda$ is bijective. Therefore, it follows from Exercise 1.5 that the spectrum of $\hat{J}_3$ is a pure point spectrum, given by $\hbar\,(\mathbb{Z} + (1/2))$ and consisting of eigenvalues of infinite multiplicity.

Solution 1.12 For $\lambda \in \mathbb{C}$, we have that

$$\begin{aligned}
&(\hat{J}_{03} - \lambda)f \\
&= {}^t\Big(\hat{L}_{03} f_1 + \frac{\hbar}{2}\,[i\sigma_3 \cdot {}^t(f_1, f_2)]_1 - \lambda f_1, \hat{L}_{03} f_2 + \frac{\hbar}{2}\,[i\sigma_3 \cdot {}^t(f_1, f_2)]_2 - \lambda f_2, \\
&\qquad \hat{L}_{03} f_3 - \frac{\hbar}{2}\,[\sigma_3 \cdot {}^t(f_3, f_4)]_1 - \lambda f_3, \hat{L}_{03} f_4 - \frac{\hbar}{2}\,[\sigma_3 \cdot {}^t(f_3, f_4)]_2 - \lambda f_4\Big) \\
&= {}^t([(\hat{J}_{L03} - \lambda)\,{}^t(f_1, f_2)]_1, [(\hat{J}_{L03} - \lambda)\,{}^t(f_1, f_2)]_2, \\
&\qquad [(\hat{J}_{R03} - \lambda)\,{}^t(f_3, f_4)]_1, [(\hat{J}_{R03} - \lambda)\,{}^t(f_3, f_4)]_2)
\end{aligned}$$

for every $f \in (D(\hat{L}_3))^4$, where $\hat{J}_{R03}$ and $\hat{J}_{L03}$ are the momenta defined by (1.21). As a consequence, $\hat{J}_{03} - \lambda$ is bijective if and only if $\hat{J}_{R03} - \lambda$ and $\hat{J}_{L03} - \lambda$ are both bijective. Therefore, it follows from Exercise 1.6 that the spectrum of $\hat{J}_{03}$ is given by $\hbar\,((i/2) + \mathbb{R}) \cup \hbar\,(-(i/2) + \mathbb{R})$.

Solution 1.13 For $\lambda \in \mathbb{C}$, we have that

$$\begin{aligned}
\left(\left(\bigtimes_{k=1}^{4} T_{a_0 v_a^0 - \vec{a}\cdot \mathrm{id}_{\mathbb{R}^3}}\right) - \lambda\right) f &= \begin{pmatrix} T_{a_0 v_a^0 - \vec{a}\cdot \mathrm{id}_{\mathbb{R}^3}} f_1 \\ T_{a_0 v_a^0 - \vec{a}\cdot \mathrm{id}_{\mathbb{R}^3}} f_2 \\ T_{a_0 v_a^0 - \vec{a}\cdot \mathrm{id}_{\mathbb{R}^3}} f_3 \\ T_{a_0 v_a^0 - \vec{a}\cdot \mathrm{id}_{\mathbb{R}^3}} f_4 \end{pmatrix} - \lambda \begin{pmatrix} f_1 \\ f_2 \\ f_3 \\ f_4 \end{pmatrix} \\
&= \begin{pmatrix} (T_{a_0 v_a^0 - \vec{a}\cdot \mathrm{id}_{\mathbb{R}^3}} - \lambda)\, f_1 \\ (T_{a_0 v_a^0 - \vec{a}\cdot \mathrm{id}_{\mathbb{R}^3}} - \lambda)\, f_2 \\ (T_{a_0 v_a^0 - \vec{a}\cdot \mathrm{id}_{\mathbb{R}^3}} - \lambda)\, f_3 \\ (T_{a_0 v_a^0 - \vec{a}\cdot \mathrm{id}_{\mathbb{R}^3}} - \lambda)\, f_4 \end{pmatrix},
\end{aligned}$$

for every $f \in \bigtimes_{k=1}^{4} D(T_{a_0 v_a^0 - \vec{a}\cdot \mathrm{id}_{\mathbb{R}^3}})$. Hence the operator $(\bigtimes_{k=1}^{4} T_{a_0 v_a^0 - \vec{a}\cdot \mathrm{id}_{\mathbb{R}^3}}) - \lambda$ is bijective if and only if λ is part of the resolvent set of $T_{a_0 v_a^0 - \vec{a}\cdot \mathrm{id}_{\mathbb{R}^3}}$. Since the spectrum of the latter operator is given by the closure of the range of $a_0 v_a^0 - \vec{a} \cdot \mathrm{id}_{\mathbb{R}^3}$, it follows that λ is part of the spectrum of $\bigtimes_{k=1}^{4} T_{a_0 v_a^0 - \vec{a}\cdot \mathrm{id}_{\mathbb{R}^3}}$ if and only if λ is part of the closure of the range of $a_0 v_a^0 - \vec{a} \cdot \mathrm{id}_{\mathbb{R}^3}$.

Solution 1.14 If $a \in \mathbb{R}^4$ is future-oriented, timelike, and, in particular, such that $a \cdot a = 1$, it follows from Corollary 5.5 of Part I the existence of a Lorentz transformation $\Lambda \in \mathcal{L}_+^{\uparrow}$ such that

$$a = \Lambda\,{}^t(1, 0, 0, 0)\ .$$

Hence, it follows for $v \in \mathbb{R}^3$ that

$$a \cdot {}^t(v_a^0(v), v_1, v_2, v_3) = (\Lambda\, {}^t(1,0,0,0)) \cdot p_a(v)$$
$$= {}^t(1,0,0,0)) \cdot (\Lambda^{-1} p_a(v)) = (\Lambda^{-1} p_a(v))_0$$
$$= (p_a(h_{a\Lambda}(v)))_0 = v_a^0(h_{a\Lambda}(v)) \ ,$$

where the parametrization $p_a : \mathbb{R}^3 \to H_{a+}$ and the bijection $h_{a\Lambda} : \mathbb{R}^3 \to \mathbb{R}^3$ are defined in Part I. Sine $h_{a\Lambda}$ is a bijection, we have that

$$\text{Ran}(a_0 v_a^0 - \vec{a} \cdot \text{id}_{\mathbb{R}^3}) = \text{Ran}(v_a^0)$$

and hence that

$$\overline{\text{Ran}(a_0 v_a^0 - \vec{a} \cdot \text{id}_{\mathbb{R}^3})} = [a, \infty) \ .$$

Solution 1.15 For $(a, G) \in \mathbb{R}^4 \times \text{SL}(2, \mathbb{C})$, $f, g \in (\mathbb{C}^{\mathbb{R}^4})^4$ and $\lambda \in \mathbb{C}$, we have that

$$R((a,G))(f+g) = \begin{pmatrix} (R_{\hbar_L}((a,G))\,{}^t((f+g)_1, (f+g)_2))_1 \\ (R_{\hbar_L}((a,G))\,{}^t((f+g)_1, (f+g)_2))_2 \\ (R_{\hbar_R}((a,G))\,{}^t((f+g)_3, (f+g)_4))_1 \\ (R_{\hbar_R}((a,G))\,{}^t((f+g)_3, (f+g)_4))_2 \end{pmatrix}$$

$$= \begin{pmatrix} (R_{\hbar_L}((a,G))\,{}^t(f_1+g_1, f_2+g_2))_1 \\ (R_{\hbar_L}((a,G))\,{}^t(f_1+g_1, f_2+g_2))_2 \\ (R_{\hbar_R}((a,G))\,{}^t(f_3+g_3, f_4+g_4))_1 \\ (R_{\hbar_R}((a,G))\,{}^t(f_3+g_3, f_4+g_4))_2 \end{pmatrix}$$

$$= \begin{pmatrix} (R_{\hbar_L}((a,G))({}^t(f_1, f_2) + {}^t(g_1, g_2)))_1 \\ (R_{\hbar_L}((a,G))({}^t(f_1, f_2) + {}^t(g_1, g_2)))_2 \\ (R_{\hbar_R}((a,G))({}^t(f_3, f_4) + {}^t(g_3, g_4)))_1 \\ (R_{\hbar_R}((a,G))({}^t(f_3, f_4) + {}^t(g_3, g_4)))_2 \end{pmatrix}$$

$$= \begin{pmatrix} (R_{\hbar_L}((a,G))\,{}^t(f_1, f_2))_1 + (R_{\hbar_L}((a,G))\,{}^t(g_1, g_2))_1 \\ (R_{\hbar_L}((a,G))\,{}^t(f_1, f_2))_2 + (R_{\hbar_L}((a,G))\,{}^t(g_1, g_2))_2 \\ (R_{\hbar_R}((a,G))\,{}^t(f_3, f_4))_1 + (R_{\hbar_R}((a,G))\,{}^t(g_3, g_4))_1 \\ (R_{\hbar_R}((a,G))\,{}^t(f_3, f_4))_2 + (R_{\hbar_R}((a,G))\,{}^t(g_3, g_4))_2 \end{pmatrix}$$

$$= \begin{pmatrix} (R_{\hbar_L}((a,G))\,{}^t(f_1, f_2))_1 \\ (R_{\hbar_L}((a,G))\,{}^t(f_1, f_2))_2 \\ (R_{\hbar_R}((a,G))\,{}^t(f_3, f_4))_1 \\ (R_{\hbar_R}((a,G))\,{}^t(f_3, f_4))_2 \end{pmatrix} + \begin{pmatrix} (R_{\hbar_L}((a,G))\,{}^t(g_1, g_2))_1 \\ (R_{\hbar_L}((a,G))\,{}^t(g_1, g_2))_2 \\ (R_{\hbar_R}((a,G))\,{}^t(g_3, g_4))_1 \\ (R_{\hbar_R}((a,G))\,{}^t(g_3, g_4))_2 \end{pmatrix}$$

$$= R((a,G))f + R((a,G))g \ ,$$

as well as that

$$R((a,G))(\lambda f) = \begin{pmatrix} (R_{\hbar_L}((a,G))\lambda\,{}^t(f_1,f_2))_1 \\ (R_{\hbar_L}((a,G))\lambda\,{}^t(f_1,f_2))_2 \\ (R_{\hbar_R}((a,G))\lambda\,{}^t(f_3,f_4))_1 \\ (R_{\hbar_R}((a,G))\lambda\,{}^t(f_3,f_4))_2 \end{pmatrix} = \lambda R((a,G))f \ .$$

Hence, $R((a,G)) \in \mathrm{Hom}((\mathbb{C}^{\mathbb{R}^4})^4, (\mathbb{C}^{\mathbb{R}^4})^4)$. Further, for (a_1,G_1), $(a_2,G_2) \in \mathbb{R}^4 \times \mathrm{SL}(2,\mathbb{C})$ and $f \in (\mathbb{C}^{\mathbb{R}^4})^4$, we have that

$$R((a_1,G_1)\cdot(a_2,G_2))f = \begin{pmatrix} (R_{\hbar_L}((a_1,G_1)\cdot(a_2,G_2))\,{}^t(f_1,f_2))_1 \\ (R_{\hbar_L}((a_1,G_1)\cdot(a_2,G_2))\,{}^t(f_1,f_2))_2 \\ (R_{\hbar_R}((a_1,G_1)\cdot(a_2,G_2))\,{}^t(f_3,f_4))_1 \\ (R_{\hbar_R}((a_1,G_1)\cdot(a_2,G_2))\,{}^t(f_3,f_4))_2 \end{pmatrix}$$

$$= \begin{pmatrix} (R_{\hbar_L}((a_1,G_1)) \circ R_{\hbar_L}((a_2,G_2))\,{}^t(f_1,f_2))_1 \\ (R_{\hbar_L}((a_1,G_1)) \circ R_{\hbar_L}((a_2,G_2))\,{}^t(f_1,f_2))_2 \\ (R_{\hbar_R}((a_1,G_1)) \circ R_{\hbar_R}((a_2,G_2))\,{}^t(f_3,f_4))_1 \\ (R_{\hbar_R}((a_1,G_1)) \circ R_{\hbar_R}((a_2,G_2))\,{}^t(f_3,f_4))_2 \end{pmatrix}$$

$$= \begin{pmatrix} (R_{\hbar_L}((a_1,G_1))(R_{\hbar_L}((a_2,G_2))\,{}^t(f_1,f_2)))_1 \\ (R_{\hbar_L}((a_1,G_1))(R_{\hbar_L}((a_2,G_2))\,{}^t(f_1,f_2)))_2 \\ (R_{\hbar_R}((a_1,G_1))(R_{\hbar_R}((a_2,G_2))\,{}^t(f_3,f_4)))_1 \\ (R_{\hbar_R}((a_1,G_1))(R_{\hbar_R}((a_2,G_2))\,{}^t(f_3,f_4)))_2 \end{pmatrix}$$

$$= R((a_1,G_1)) \begin{pmatrix} ((R_{\hbar_L}((a_2,G_2))\,{}^t(f_1,f_2)))_1 \\ ((R_{\hbar_L}((a_2,G_2))\,{}^t(f_1,f_2)))_2 \\ = ((R_{\hbar_R}((a_2,G_2))\,{}^t(f_3,f_4)))_1 \\ ((R_{\hbar_R}((a_2,G_2))\,{}^t(f_3,f_4)))_2 \end{pmatrix}$$

$$= R((a_1,G_1))(R((a_2,G_2))f) = [R((a_1,G_1)) \circ R((a_2,G_2))]f \ ,$$

and

$$R((0,E))f := \begin{pmatrix} (R_{\hbar_L}((0,E))\,{}^t(f_1,f_2))_1 \\ (R_{\hbar_L}((0,E))\,{}^t(f_1,f_2))_2 \\ (R_{\hbar_R}((0,E))\,{}^t(f_3,f_4))_1 \\ (R_{\hbar_R}((0,E))\,{}^t(f_3,f_4))_2 \end{pmatrix} = \begin{pmatrix} ({}^t(f_1,f_2))_1 \\ ({}^t(f_1,f_2))_2 \\ ({}^t(f_3,f_4))_1 \\ ({}^t(f_3,f_4))_2 \end{pmatrix}$$

$$= \begin{pmatrix} f_1 \\ f_2 \\ f_3 \\ f_4 \end{pmatrix} = f \ .$$

Hence, it follows that

$$R((a_1,G_1)\cdot(a_2,G_2)) = R((a_1,G_1)) \circ R((a_2,G_2)) \ ,$$
$$R((0,E)) = \mathrm{id}_{(\mathbb{C}^{\mathbb{R}^4})^4}$$

and that the map $R : \mathbb{R}^4 \times \mathrm{SL}(2, \mathbb{C}) \to \mathrm{Hom}((\mathbb{C}^{\mathbb{R}^4})^4, (\mathbb{C}^{\mathbb{R}^4})^4)$, that associates with every $(a, G) \in \mathbb{R}^4 \times \mathrm{SL}(2, \mathbb{C})$ the map $R((a, G))$, is a representation of $\mathbb{R}^4 \times \mathrm{SL}(2, \mathbb{C})$. Finally, since, $R_{\hbar_L}, R_{\hbar_R}$ leave $(C^1(\mathbb{R}^4, \mathbb{C}))^2$ invariant, it follows that R leaves $(C^1(\mathbb{R}^4, \mathbb{C}))^4$ invariant.

Solution 2.1 For $F_1, G_1, F_2, G_2 \in \mathrm{Skew}(4, \mathbb{C})$ and $\lambda \in \mathbb{C}$, we have that

$$
\begin{aligned}
&\langle F_1 + G_1, F_2\rangle = \frac{1}{2}.\mathrm{Tr}((F_1 + G_1)^* F_2) = \frac{1}{2}.\mathrm{Tr}((F_1^* + G_1^*)F_2) \\
&= \frac{1}{2}.\mathrm{Tr}(F_1^* F_2 + G_1^* F_2) = \frac{1}{2}.\mathrm{Tr}(F_1^* F_2) + \frac{1}{2}.\mathrm{Tr}(G_1^* F_2) = \langle F_1, F_2\rangle + \langle G_1, F_2\rangle \ , \\
&\langle \lambda.F_1, F_2\rangle = \frac{1}{2}.\mathrm{Tr}((\lambda F_1)^* F_2) = \frac{1}{2}.\mathrm{Tr}((\lambda^* F_1^*)F_2) = \frac{1}{2}.\mathrm{Tr}(\lambda^*(F_1^* F_2)) \\
&= \lambda^* \frac{1}{2}.\mathrm{Tr}(F_1^* F_2) = \lambda^* \langle F_1, F_2\rangle \ , \\
&\langle F_1, (F_2 + G_2)\rangle = \frac{1}{2}.\mathrm{Tr}(F_1^*(F_2 + G_2)) = \frac{1}{2}.\mathrm{Tr}(F_1^* F_2 + F_1^* G_2) \\
&= \frac{1}{2}.\mathrm{Tr}(F_1^* F_2) + \frac{1}{2}.\mathrm{Tr}(F_1^* G_2) = \langle F_1, F_2\rangle + \langle F_1, G_2\rangle \ , \\
&\langle F_1, \lambda.F_2\rangle = \frac{1}{2}.\mathrm{Tr}(F_1^*(\lambda F_2)) = \frac{1}{2}.\mathrm{Tr}(\lambda(F_1^* F_2)) = \lambda \frac{1}{2}.\mathrm{Tr}(F_1^* F_2) \\
&= \lambda\langle F_1, F_2\rangle \ , \\
&\langle F_2, F_1\rangle = \frac{1}{2}.\mathrm{Tr}(F_2^* F_1) = \frac{1}{2}.\mathrm{Tr}((F_1^* F_2)^*) = \left(\frac{1}{2}.\mathrm{Tr}(F_1^* F_2)\right)^* = \langle F_1, F_2\rangle^* \ .
\end{aligned}
$$

Hence, by

$$\langle F, G\rangle := \frac{1}{2}.\mathrm{Tr}(F^* G) \ ,$$

for ever $F, G \in \mathrm{Skew}(4, \mathbb{C})$, there is defined a Hermitian $\mathbb{C}$-Sesquilinear form $\langle\ ,\ \rangle : \mathrm{Skew}(4, \mathbb{C}) \times \mathrm{Skew}(4, \mathbb{C}) \to \mathbb{C}$ on $\mathrm{Skew}(4, \mathbb{C})$. Further, a short calculation shows that

$$\frac{1}{2}.\mathrm{Tr}(F^* G) = F_{01}^* G_{01} + F_{02}^* G_{02} + F_{03}^* G_{03} + F_{32}^* + F_{13}^* G_{13} + F_{21}^* G_{21}$$

and hence that

$$\langle F, F\rangle = \frac{1}{2}.\mathrm{Tr}(F^* F) = |F_{01}|^2 + |F_{02}|^2 + |F_{03}|^2 + |F_{32}|^2 + |F_{13}|^2 + |F_{21}|^2 \ ,$$

for

$$F = \begin{pmatrix} 0 & F_{01} & F_{02} & F_{03} \\ -F_{01} & 0 & -F_{21} & F_{13} \\ -F_{02} & F_{21} & 0 & -F_{32} \\ -F_{03} & -F_{13} & F_{32} & 0 \end{pmatrix} , \; G = \begin{pmatrix} 0 & G_{01} & G_{02} & G_{03} \\ -G_{01} & 0 & -G_{21} & G_{13} \\ -G_{02} & G_{21} & 0 & -G_{32} \\ -G_{03} & -G_{13} & G_{32} & 0 \end{pmatrix}$$
$$\in \text{Skew}(4, \mathbb{C}) \; .$$

Since, for $F \in \text{Skew}(4, \mathbb{C})\setminus\{0\}$, $\langle F, F\rangle > 0$, $\langle \; , \; \rangle$ is a scalar product.

Solution 2.2 For

$$F = \begin{pmatrix} 0 & F_{01} & F_{02} & F_{03} \\ -F_{01} & 0 & -F_{21} & F_{13} \\ -F_{02} & F_{21} & 0 & -F_{32} \\ -F_{03} & -F_{13} & F_{32} & 0 \end{pmatrix} \in \text{Skew}(4, \mathbb{C}) \; ,$$

we have that

$$|F|_{\text{Tr}}^2 = |F_{01}|^2 + |F_{02}|^2 + |F_{03}|^2 + |F_{32}|^2 + |F_{13}|^2 + |F_{21}|^2 \leqslant 6 \, \|F\|_\infty^2$$

and hence that

$$|F|_{\text{Tr}} \leqslant \sqrt{6} \, \|F\|_\infty \; .$$

Further,

$$\|F\|_\infty = \max\{|F_{01}|, |F_{02}|, |F_{03}|, |F_{32}|, |F_{13}|, |F_{21}|\} \leqslant |F|_{\text{Tr}} \; .$$

Therefore, it follows that

$$|F|_{\text{Tr}} \leqslant \sqrt{6} \, \|F\|_\infty \leqslant \sqrt{6} \, |F|_{\text{Tr}} \; .$$

Solution 2.3 For $S_1, S_2 \in \text{GL}(4, \mathbb{R})$, $F \in \text{Skew}(4, \mathbb{C})$, it follows that

$$\begin{aligned}
&(S_1^{-1})^t \cdot F \cdot S_1^{-1} - (S_2^{-1})^t \cdot F \cdot S_2^{-1} \\
&= (S_1^{-1})^t \cdot F \cdot S_1^{-1} - (S_1^{-1})^t \cdot F \cdot S_2^{-1} + (S_1^{-1})^t \cdot F \cdot S_2^{-1} - (S_2^{-1})^t \cdot F \cdot S_2^{-1} \\
&= (S_1^{-1})^t \cdot F \cdot (S_1^{-1} - S_2^{-1}) + (S_1^{-1} - S_2^{-1})^t \cdot F \cdot S_2^{-1} \\
&= (S_1^{-1})^t \cdot F \cdot (S_1^{-1} - S_2^{-1}) + (S_1^{-1} - S_2^{-1})^t \cdot F \cdot (S_2^{-1} - S_1^{-1}) \\
&\quad + (S_1^{-1} - S_2^{-1})^t \cdot F \cdot S_1^{-1} \; .
\end{aligned}$$

Further, since for $A, B \in \text{M}(n, \mathbb{K})$, where $n \in \mathbb{N}^*$ and $\mathbb{K} \in \{\mathbb{R}, \mathbb{C}\}$,

$$|(A \cdot B)_{jk}| = |\sum_{l=1}^{n} A_{jl} B_{lk}| \leqslant \sum_{l=1}^{n} |A_{jl}| \cdot |B_{lk}| \leqslant n \, \|A\|_\infty \cdot \|B\|_\infty \, ,$$

and hence

$$\|A \cdot B\|_\infty \leqslant n \, \|A\|_\infty \cdot \|B\|_\infty \, ,$$

it follows that

$$\begin{aligned} \|\mathfrak{R}(S_2)F - \mathfrak{R}(S_1)F\|_\infty &= \|(S_2^{-1})^t \cdot F \cdot S_2^{-1} - (S_1^{-1})^t \cdot F \cdot S_1^{-1}\|_\infty \\ &\leqslant 32 \, \|(S_1^{-1})^t\|_\infty \cdot \|F\|_\infty \cdot \|S_2^{-1} - S_1^{-1}\|_\infty + 16 \, \|F\|_\infty \cdot \|S_2^{-1} - S_1^{-1}\|_\infty^2 \, . \end{aligned}$$

Further, using the equivalence of the norms $\| \; \|_\infty$ and $| \; |_{\mathrm{Tr}}$ from Exercise 2.2, it follows that

$$\begin{aligned} &\|\mathfrak{R}(S_2)F - \mathfrak{R}(S_1)F\|_{\mathrm{Tr}} \\ &\leqslant 16 \sqrt{6} \, (2 \, \|(S_1^{-1})^t\|_\infty \|S_2^{-1} - S_1^{-1}\|_\infty + \|S_2^{-1} - S_1^{-1}\|_\infty^2) \, \|F\|_{\mathrm{Tr}} \end{aligned}$$

and hence that

$$\|\mathfrak{R}(S_2) - \mathfrak{R}(S_1)\|_{\mathrm{op}} \leqslant 16 \sqrt{6} \, (2 \, \|(S_1^{-1})^t\|_\infty \|S_2^{-1} - S_1^{-1}\|_\infty + \|S_2^{-1} - S_1^{-1}\|_\infty^2) \, .$$

Since, $\mathrm{GL}(4, \mathbb{R})$ is a topological group, i.e., the operations multiplication and inversion are continuous maps, see Sect. 3.1.2 of Part I, and the norms $\| \; \|_{\mathrm{op}}$ and $\| \; \|_\infty$ on $\mathrm{GL}(4, \mathbb{R})$ are equivalent, see Inequality 3.1 in Part I, it follows that $\mathfrak{R}$ is continuous.

Solution 2.4 For every $\psi \in \mathbb{C}^6$, we have that $\mathfrak{I}(\psi) \in \mathrm{Skew}(4, \mathbb{C})$ and hence that the map $\mathfrak{I} : \mathbb{C}^6 \to \mathrm{Skew}(4, \mathbb{C})$ is well-defined. Further, $\mathfrak{I}$ is obviously linear, with a vanishing kernel and hence injective. Further, for $F \in \mathrm{Skew}(4, \mathbb{C})$, we have that

$$\mathfrak{I} \begin{pmatrix} F_{01} \\ F_{02} \\ F_{03} \\ F_{32} \\ F_{13} \\ F_{21} \end{pmatrix} = \begin{pmatrix} 0 & F_{01} & F_{02} & F_{03} \\ -F_{01} & 0 & -F_{21} & F_{13} \\ -F_{02} & F_{21} & 0 & -F_{32} \\ -F_{03} & -F_{13} & F_{32} & 0 \end{pmatrix} = F$$

and hence that $\mathfrak{I}$ is surjective as well that

$$\mathfrak{I}^{-1}(F) = \begin{pmatrix} F_{01} \\ F_{02} \\ F_{03} \\ F_{32} \\ F_{13} \\ F_{21} \end{pmatrix} \, .$$

In addition, we have that

$$\|\mathfrak{I}(\psi)\|_{\mathrm{Tr}}^2 = \left\| \begin{pmatrix} 0 & \psi_1 & \psi_2 & \psi_3 \\ -\psi_1 & 0 & -\psi_6 & \psi_5 \\ -\psi_2 & \psi_6 & 0 & -\psi_4 \\ -\psi_3 & -\psi_5 & \psi_4 & 0 \end{pmatrix} \right\|_{\mathrm{Tr}}^2 = \sum_{j=1}^{6} |\psi_j|^2 = |\psi|^2 ,$$

for every $\psi \in \mathbb{C}^6$. Hence, $\mathfrak{I}$ is isometric and therefore also a Hilbert space isomorphism.

Solution 2.5 If $j \in \{1, 2, 3\}$, $s \in \mathbb{R}$, then the no-zero diagonal components of $D_{\mathfrak{M}}(M_j(s)) - 1$ are given by $\cos(s) - 1$, whereas the non-zero non-diagonal components are given by $\pm\sin(s)$. Since

$$|\cos(s) - 1| = |\cos(|s|) - 1| = |\int_0^{|s|} \sin(\tau)\, d\tau| \leqslant \int_0^{|s|} |\sin(\tau)|\, d\tau \leqslant \int_0^{|s|} d\tau$$
$$= |s| \ , \ |\sin(s)| = \sin(|s|) \leqslant |s| \ ,$$

we have that

$$\|D_{\mathfrak{M}}(M_j(s)) - 1\|_\infty \leqslant |s| \ .$$

Solution 2.6 For $s \in \mathbb{R}^*$, we have that

$$\frac{1}{s}\,(D_{\mathfrak{M}}(M_1(s)) - 1) - \begin{pmatrix} 0 & 0 & 0 & 0 & 0 & 0 \\ 0 & 0 & 1 & 0 & 0 & 0 \\ 0 & -1 & 0 & 0 & 0 & 0 \\ 0 & 0 & 0 & 0 & 0 & 0 \\ 0 & 0 & 0 & 0 & 0 & 1 \\ 0 & 0 & 0 & 0 & -1 & 0 \end{pmatrix}$$
$$= \begin{pmatrix} 0 & 0 & 0 & 0 & 0 & 0 \\ 0 & \frac{\cos(s)-1}{s} & \frac{\sin(s)-s}{s} & 0 & 0 & 0 \\ 0 & \frac{s-\sin(s)}{s} & \frac{\cos(s)-1}{s} & 0 & 0 & 0 \\ 0 & 0 & 0 & 0 & 0 & 0 \\ 0 & 0 & 0 & 0 & \frac{\cos(s)-1}{s} & \frac{\sin(s)-s}{s} \\ 0 & 0 & 0 & 0 & \frac{s-\sin(s)}{s} & \frac{\cos(s)-1}{s} \end{pmatrix} ,$$

$$\frac{1}{s}\,(D_{\mathfrak{M}}(M_2(s)) - 1) - \begin{pmatrix} 0 & 0 & -1 & 0 & 0 & 0 \\ 0 & 0 & 0 & 0 & 0 & 0 \\ 1 & 0 & 0 & 0 & 0 & 0 \\ 0 & 0 & 0 & 0 & 0 & -1 \\ 0 & 0 & 0 & 0 & 0 & 0 \\ 0 & 0 & 0 & 1 & 0 & 0 \end{pmatrix}$$

$$= \begin{pmatrix} \frac{\cos(s)-1}{s} & 0 & \frac{s-\sin(s)}{s} & 0 & 0 & 0 \\ 0 & 0 & 0 & 0 & 0 & 0 \\ \frac{\sin(s)-s}{s} & 0 & \frac{\cos(s)-1}{s} & 0 & 0 & 0 \\ 0 & 0 & 0 & \frac{\cos(s)-1}{s} & 0 & \frac{s-\sin(s)}{s} \\ 0 & 0 & 0 & 0 & 0 & 0 \\ 0 & 0 & 0 & \frac{\sin(s)-s}{s} & 0 & \frac{\cos(s)-1}{s} \end{pmatrix},$$

$$\frac{1}{s}\left(D_{\mathfrak{M}}(M_3(s)) - 1\right) - \begin{pmatrix} 0 & 1 & 0 & 0 & 0 & 0 \\ -1 & 0 & 0 & 0 & 0 & 0 \\ 0 & 0 & 0 & 0 & 0 & 0 \\ 0 & 0 & 0 & 0 & 1 & 0 \\ 0 & 0 & 0 & -1 & 0 & 0 \\ 0 & 0 & 0 & 0 & 0 & 0 \end{pmatrix}$$

$$= \begin{pmatrix} \frac{\cos(s)-1}{s} & \frac{\sin(s)-s}{s} & 0 & 0 & 0 & 0 \\ \frac{s-\sin(s)}{s} & \frac{\cos(s)-1}{s} & 0 & 0 & 0 & 0 \\ 0 & 0 & 0 & 0 & 0 & 0 \\ 0 & 0 & 0 & \frac{\cos(s)-1}{s} & \frac{\sin(s)-s}{s} & 0 \\ 0 & 0 & 0 & \frac{s-\sin(s)}{s} & \frac{\cos(s)-1}{s} & 0 \\ 0 & 0 & 0 & 0 & 0 & 0 \end{pmatrix}.$$

Further,

$$\left|\frac{1-\cos(s)}{s}\right| = \frac{1-\cos(|s|)}{|s|} = \frac{1}{|s|}\int_0^{|s|} \sin(\tau)\, d\tau \leqslant \frac{1}{|s|}\int_0^{|s|} |\sin(\tau)|\, d\tau$$
$$\leqslant \frac{1}{|s|}\int_0^{|s|} \tau\, d\tau = \frac{|s|}{2},$$
$$\left|\frac{s-\sin(s)}{s}\right| = \left|\frac{1}{s}\int_0^{s} [1-\cos(\tau)]\, d\tau\right| = \frac{1}{|s|}\int_0^{|s|} [1-\cos(\tau)]\, d\tau$$
$$\leqslant \frac{1}{|s|}\int_0^{|s|} \frac{\tau^2}{2}\, d\tau = \frac{1}{|s|}\frac{|s|^3}{6} = \frac{|s|^2}{6}.$$

Hence, it follows for $j \in \{1, 2, 3\}$, $s \in [-1, 1]$ that

$$\|\frac{1}{s}\left[D_{\mathfrak{M}}(M_j(s)) - 1\right] - i\hat{\sigma}_j\|_\infty \leqslant \frac{|s|}{2}.$$

Solution 2.7 If $j \in \{1, 2, 3\}$, $s \in [-1, 1]$, then the no-zero diagonal components of $D_{\mathfrak{M}}(M_{0j}(s)) - 1$, $j \in \{1, 2, 3\}$, are given by $\cosh(s) - 1$, whereas the non-zero non-diagonal components are given by $\pm\sinh(s)$.

$$|\cosh(s)-1| = |\cosh(|s|)-1| = |\int_0^{|s|} \sinh(\tau)\,d\tau| = \int_0^{|s|} \sinh(\tau)\,d\tau$$
$$\leqslant \int_0^{|s|} \sinh(|s|)\,d\tau = |s|\sinh(|s|) \leqslant \sinh(|s|) \ ,$$

we have that

$$\|D_{\mathfrak{M}}(M_{0j}(s)) - 1\|_\infty \leqslant \sinh(|s|) \ .$$

Solution 2.8 For $s \in \mathbb{R}^*$, we have that

$$\frac{1}{s}\left[D_{\mathfrak{M}}(M_{01}(s)) - 1\right] - \begin{pmatrix} 0&0&0&0&0&0\\ 0&0&0&0&0&1\\ 0&0&0&0&-1&0\\ 0&0&0&0&0&0\\ 0&0&-1&0&0&0\\ 0&1&0&0&0&0 \end{pmatrix}$$
$$= \begin{pmatrix} 0&0&0&0&0&0\\ 0&\frac{\cosh(s)-1}{s}&0&0&0&\frac{\sinh(s)-s}{s}\\ 0&0&\frac{\cosh(s)-1}{s}&0&\frac{s-\sinh(s)}{s}&0\\ 0&0&0&0&0&0\\ 0&0&\frac{s-\sinh(s)}{s}&0&\frac{\cosh(s)-1}{s}&0\\ 0&\frac{\sinh(s)-s}{s}&0&0&0&\frac{\cosh(s)-1}{s} \end{pmatrix},$$

$$\frac{1}{s}\left[D_{\mathfrak{M}}(M_{02}(s)) - 1\right] - \begin{pmatrix} 0&0&0&0&0&-1\\ 0&0&0&0&0&0\\ 0&0&0&1&0&0\\ 0&0&1&0&0&0\\ 0&0&0&0&0&0\\ -1&0&0&0&0&0 \end{pmatrix}$$
$$= \begin{pmatrix} \frac{\cosh(s)-1}{s}&0&0&0&0&\frac{s-\sinh(s)}{s}\\ 0&0&0&0&0&0\\ 0&0&\frac{\cosh(s)-1}{s}&\frac{\sinh(s)-s}{s}&0&0\\ 0&0&\frac{\sinh(s)-s}{s}&\frac{\cosh(s)-1}{s}&0&0\\ 0&0&0&0&0&0\\ \frac{s-\sinh(s)}{s}&0&0&0&0&\frac{\cosh(s)-1}{s} \end{pmatrix},$$

$$\frac{1}{s}\left[D_{\mathfrak{M}}(M_{03}(s)) - 1\right] - \begin{pmatrix} 0&0&0&0&1&0\\ 0&0&0&-1&0&0\\ 0&0&0&0&0&0\\ 0&-1&0&0&0&0\\ 1&0&0&0&0&0\\ 0&0&0&0&0&0 \end{pmatrix}$$

$$= \begin{pmatrix} \frac{\cosh(s)-1}{s} & 0 & 0 & 0 & \frac{\sinh(s)-s}{s} & 0 \\ 0 & \frac{\cosh(s)-1}{s} & 0 & \frac{s-\sinh(s)}{s} & 0 & 0 \\ 0 & 0 & 0 & 0 & 0 & 0 \\ 0 & \frac{s-\sinh(s)}{s} & 0 & \frac{\cosh(s)-1}{s} & 0 & 0 \\ \frac{\sinh(s)-s}{s} & 0 & 0 & 0 & \frac{\cosh(s)-1}{s} & 0 \\ 0 & 0 & 0 & 0 & 0 & 0 \end{pmatrix} .$$

Further,

$$\left| \frac{\cosh(s) - 1}{s} \right| = \frac{\cosh(|s|) - 1}{|s|} = \frac{1}{|s|} \int_0^{|s|} \sinh(\tau)\, d\tau$$
$$\leqslant \frac{1}{|s|} \int_0^{|s|} \sinh(|s|)\, d\tau = \sinh(|s|) \ ,$$
$$\left| \frac{\sinh(s) - s}{s} \right| = \frac{1}{|s|} \, | \sinh(|s|) - |s| \, | = \frac{1}{|s|} \int_0^{|s|} [\cosh(\tau) - 1]\, d\tau$$
$$\leqslant \frac{1}{|s|} \int_0^{|s|} \tau \sinh(\tau)\, d\tau \leqslant \frac{1}{|s|} \sinh(|s|) \int_0^{|s|} \tau\, d\tau = \frac{|s|}{2} \sinh(|s|) \ .$$

Hence, it follows for $j \in \{1, 2, 3\}$, $s \in [-1, 1]$ that

$$\| \frac{1}{is} \left[D_{\mathfrak{M}}(M_{0j}(s)) - 1 \right] - \hat{\sigma}_j \|_\infty \leqslant \sinh(|s|) \ .$$

Solution 2.9 For $\lambda \in \mathbb{C}$, we have that

$$\left(\bigtimes_{k=1}^{6} T_{a_0 v_a^0 - \vec{a} \cdot \mathrm{id}_{\mathbb{R}^3}} - \lambda \right) f = \begin{pmatrix} T_{a_0 v_a^0 - \vec{a} \cdot \mathrm{id}_{\mathbb{R}^3}} f_1 \\ \vdots \\ T_{a_0 v_a^0 - \vec{a} \cdot \mathrm{id}_{\mathbb{R}^3}} f_6 \end{pmatrix} - \lambda \begin{pmatrix} f_1 \\ \vdots \\ f_6 \end{pmatrix}$$
$$= \begin{pmatrix} (T_{a_0 v_a^0 - \vec{a} \cdot \mathrm{id}_{\mathbb{R}^3}} - \lambda) f_1 \\ \vdots \\ (T_{a_0 v_a^0 - \vec{a} \cdot \mathrm{id}_{\mathbb{R}^3}} - \lambda) f_6 \end{pmatrix} ,$$

for every $f \in \bigtimes_{k=1}^{6} D(T_{a_0 v_a^0 - \vec{a} \cdot \mathrm{id}_{\mathbb{R}^3}})$. Hence the operator $(\bigtimes_{k=1}^{6} T_{a_0 v_a^0 - \vec{a} \cdot \mathrm{id}_{\mathbb{R}^3}}) - \lambda$ is bijective if and only if λ is part of the resolvent set of $T_{a_0 v_a^0 - \vec{a} \cdot \mathrm{id}_{\mathbb{R}^3}}$. Since the spectrum of the latter operator is given by the closure of the range of $a_0 v_a^0 - \vec{a} \cdot \mathrm{id}_{\mathbb{R}^3}$, it follows that λ is part of the spectrum of $\bigtimes_{k=1}^{6} T_{a_0 v_a^0 - \vec{a} \cdot \mathrm{id}_{\mathbb{R}^3}}$ if and only if λ is part of the closure of the range of $a_0 v_a^0 - \vec{a} \cdot \mathrm{id}_{\mathbb{R}^3}$.

Solution 2.10 If $a \in \mathbb{R}^4$ is future-oriented, timelike, and, in particular, such that $a \cdot a = 1$, it follows from Corollary 5.5 of Part I the existence of a Lorentz transformation $\Lambda \in \mathcal{L}_+^\uparrow$ such that

$$a = \Lambda\,^t(1, 0, 0, 0)\ .$$

Hence, it follows for $v \in \mathbb{R}^3$ that

$$\begin{aligned} a \cdot {}^t(v_a^0(v), v_1, v_2, v_3) &= (\Lambda\,^t(1, 0, 0, 0)) \cdot p_a(v) \\ &= {}^t(1, 0, 0, 0)) \cdot (\Lambda^{-1} p_a(v)) = (\Lambda^{-1} p_a(v))_0 \\ &= (p_a(h_{a\Lambda}(v)))_0 = v_a^0(h_{a\Lambda}(v))\ , \end{aligned}$$

where the parametrization $p_a : \mathbb{R}^3 \to H_{a+}$ and the bijection $h_{a\Lambda} : \mathbb{R}^3 \to \mathbb{R}^3$ are defined in Part I. Sine $h_{a\Lambda}$ is a bijecion, we have that

$$\mathrm{Ran}(a_0 v_a^0 - \vec{a} \cdot \mathrm{id}_{\mathbb{R}^3}) = \mathrm{Ran}(v_a^0)$$

and hence that

$$\overline{\mathrm{Ran}(a_0 v_a^0 - \vec{a} \cdot \mathrm{id}_{\mathbb{R}^3})} = [a, \infty)\ .$$

References

1. Baez JC, Segal IE, Zhou Z (1992) Introduction to algebraic and constructive quantum field theory. Princeton, Princeton University Press
2. Bargmann V, Wigner EP (1948) Group theoretical discussion of relativistic wave equations. Proc Natl Acad Sci USA, series 34:211–223
3. Barut AO, Raczka R (1980) Theory of group representations and applications, 2nd edn. Polish Scientific Publishers, Warszawa
4. Beyer HR (2007) Beyond partial differential equations: On linear and quasi-linear abstract hyperbolic evolution equations, vol 1898. Lecture Notes in Mathematics, Springer, Berlin
5. Beyer HR (1991) Remarks on Fulling's quantization. Class. Quantum Grav. series 8:1091–1112
6. Beyer H (2022) The reasoning of quantum mechanics: Operator theory and the harmonic oscillator. Cham, Springer Nature
7. Beyer H (2023) Introduction to quantum mechanics: Physics and Operator Theory. Cham, Springer Nature
8. Beyer HR (2010) Calculus and analysis: A combined approach. Wiley, New York
9. Bjorken JD, Drell SD (1964) Relativistic quantum mechanics. McGraw-Hill, New York
10. Bjorken JD, Drell SD (1965) Relativistic quantum fields. McGraw-Hill, New York
11. Brezis H (1983) Analyse fonctionnelle: Théorie et applications. Collection Mathématiques Appliquées pour la Maîtrise, Masson, Paris
12. Bonanos S 2003, *Capabilities of the Mathematica package 'Riemannian Geometry and Tensor Calculus'*, Recent Developments in Gravity, 174–182
13. Buchholz D 2000, *Algebraic quantum field theory: A status report*, Plenary talk given at XIIIth International Congress on Mathematical Physics, London, http://xxx.lanl.gov/abs/math-ph/0011044
14. Carmeli M, Malin M (2000) Theory of spinors: An introduction. World Scientific, Singapore
15. Davydov AS (1965) Quantum mechanics. Pergamon press, Oxford
16. Dirac PAM (1928) The quantum theory of the electron. Proceedings of the Royal Society of London, Series A 117:610–624
17. Dixmier J (1977) $\mathbb{C}^*$ -Algebras. North-Holland, Amsterdam

H. R. Beyer, *Quantum Spin and Representations of the Poincaré Group, Part II*, Synthesis Lectures on Engineering, Science, and Technology,
https://doi.org/10.1007/978-3-031-95823-6

18. Dunford N, Schwartz JT (1957) Linear operators, Part I: General theory. Wiley, New York
19. Dunford N, Schwartz JT (1963) Linear operators, Part II: Spectral theory: Self adjoint operators in Hilbert space theory. Wiley, New York
20. Engel K-J, Nagel R (2000) One-parameter semigroups for linear evolution equations. Springer, New York
21. Fulling SA (1989) Aspects of quantum field theory in curved spacetime. Cambridge University Press, Cambridge
22. Gelfand IM, Naimark MA (1957) Unitäre Darstellungen der klassischen Gruppen. Akademie-Verlag, Berlin
23. Goldberg S 1985, *Unbounded linear operators* Dover: New York
24. Goldstein JA (1985) Semigroups of linear operators and applications. Oxford University Press, New York
25. Gromoll D, Klingenberg W, Meyer W (1975) Riemannsche Geometrie im Großen. Springer, Berlin
26. Haag R (1996) Local quantum physics: Fields, particles, algebras. Springer, New York
27. Hamermesh M (1989) Group theory and its applications to physical problems. Dover, New York
28. Helgason S (1978) Differential geometry, Lie groups, and symmetric spaces. Academic Press, New York
29. Hille E, Phillips RS (1957) Functional analysis and semi-groups. Revised, Providence, AMS
30. Hirzebruch F, Scharlau W (1971) Einführung in die Funktionalanalysis. Mannheim, BI
31. Hladik J (1999) Spinors in physics. Springer, New York
32. Kaku M (1993) Quantum field theory: A modern introduction. Oxford University Press, New York
33. Kato T (1966) Perturbation theory for linear operators. Springer, New York
34. Lang S (1972) Differentiable manifolds. Addison-Wesley, Reading
35. Lang S 1996, *Real and functional analysis*, 3rd ed., Springer: New York
36. Lang S (1997) Undergraduate analysis, 2nd edn. Springer, New York
37. Mackey GW (1951) On Induced representations of groups. American Journal of Mathematics, series 73:576–592
38. Mackey GW (2004) Mathematical foundations of quantum mechanics. Dover, Dover Publications
39. Maggiore M (2005) A Modern introduction to quantum field theory. Oxford University Press, Oxford
40. Messiah A (2014) Quantum mechanics. Dover Publication, New York
41. Von Neumann J (1930) Allgemeine Eigenwerttheorie hermitescher Funktionaloperatoren. Mathematische Annalen 102:49–131
42. Von Neumann J (1932) Mathematische Grundlagen der Quantenmechanik. Springer, Berlin
43. Ohnuki Y (1988) Unitary representations of the Poincaré group and relativistic wave equations. World Scientific, Singapore
44. Pazy A (1983) Semigroups of linear operators and applications to partial differential equations. Springer, New York
45. Penrose R, Rindler W (1987) Spinors and space-time, vol 1. Cambridge University Press, Cambridge
46. Peskin ME, Schroeder DV (2018) An Introduction to quantum field theory. CRC Press, Boca Raton
47. Prugovecki E (1981) Quantum mechanics in Hilbert space. Academic Press, New York
48. Ratcliffe J G 2019, *Foundations of hyperbolic manifolds*, 3rd ed., Springer: New York
49. Reed M and Simon B, 1980, 1975, 1979, 1978, *Methods of modern mathematical physics*, Volume I, II, III, IV, Academic: New York

50. Riesz F, Sz-Nagy B (1955) Functional analysis. Unger, New York
51. Rudin W (1991) Functional analysis, 2nd edn. MacGraw-Hill, New York
52. Ryder LH (1996) Quantum field theory, 2nd edn. Cambridge University Press, Cambridge
53. Sakurai JJ (1967) Advanced quantum mechanics. Addison-Wesley, Reading
54. Schechter M (2003) Operator methods in quantum mechanics. Dover Publication, New York
55. Schiff LI (1968) Quantum mechanics, 3Rev edn. McGraw-Hill Education, New York
56. Schroer B 2001, *Lectures on algebraic quantum field theory and operator algebras*, http://xxx.lanl.gov/abs/math-ph/0102018
57. Simon B (2015) A comprehensive course in analysis Part 4: Operator theory. Providence, AMS
58. Sternberg S (1995) Group theory and physics. Cambridge University Press, Cambridge
59. Streater RF, Wightman AS (2000) PCT, spin and statistics, and all that. Princeton, Princeton University Press
60. Thirring W (1981) A course in mathematical physics 3: Quantum Mechanics of Atoms and Molecules. Springer, New York
61. Thorpe JA (1979) Elementary topics in differential geometry. Springer, New York
62. Tomonaga S (1997) The story of spin. The university of Chicago press, Chicago
63. Tung W-K (1985) Group theory in physics. World Scientific, Singapore
64. van der Waerden BL (1932) Die gruppentheoretische Methode in der Quantenmechanik. Springer, Berlin
65. Wald RM (1994) Quantum field theory in curved spacetime and black hole thermodynamics. University of Chicago Press, Chicago
66. Warner FW (2010) Foundations of differentiable manifolds and Lie groups. Springer, New York
67. Weidmann J (1980) Linear operators in Hilbert spaces. Springer, New York
68. Weidmann J (2000) Lineare Operatoren in Hilberträumen: Teil I: Grundlagen. Teubner, Stuttgart
69. Weidmann J (2003) Lineare Operatoren in Hilberträumen: Teil II: Anwendungen. Teubner, Stuttgart
70. Weinberg S (1995) The quantum theory of fields, vol I. University of Cambridge, Cambridge
71. Weyl H (1927) Quantenmechanik und Gruppentheorie. Zeitschrift für Physik 46:1–46
72. Weyl H (1929) Elektron und Gravitation I. Zeitschrift für Physik 56:330–352
73. Wigner E (1939) On unitary representations of the inhomogeneous Lorentz group. Annals of Mathematics 40:149–204
74. Yosida K (1968) Functional analysis, 2nd edn. Springer, Berlin

Index

H. R. Beyer, *Quantum Spin and Representations of the Poincaré Group, Part II*, Synthesis Lectures on Engineering, Science, and Technology,
https://doi.org/10.1007/978-3-031-95823-6

www.ingramcontent.com/pod-product-compliance
Lightning Source LLC
Chambersburg PA
CBHW080817020925
31944CB00005B/253
* 9 7 8 3 0 3 1 9 5 8 2 2 9 *